Science versus practice

Robert Bud and Gerrylynn K. Roberts

Science versus practice

Chemistry in Victorian Britain

Manchester University Press

Copyright © Robert Bud and Gerrylynn K. Roberts 1984

Published by Manchester University Press, Oxford Road, Manchester, M13 9PL
51 Washington Street, Dover, New Hampshire 03820, USA

British Library cataloguing in publication data

Bud, Robert
 Science versus practice.
 1. Chemistry—Great Britain—History
 19th century
 I. Title   II. Roberts, Gerrylynn K.
 540'.941   QD18.G7

Library of Congress cataloging in publication data

Bud, Robert.
 Science versus practice.
 Includes bibliographical references and index.
 1. Chemistry—Great Britain—History.   I. Roberts, Gerrylynn K.   II Title.
 QD18.G7B83   1984   540'.941   84–853

ISBN 0–7190–1070–5

Phototypesetting by Saxon Press, Derby

Printed in Great Britain by
Butler & Tanner Ltd, Frome and London

Contents

Preface

This project grew out of our separate experiences of research into various aspects of Victorian chemistry. However, this book, both in writing and in research, is a joint endeavour that has gone beyond our earlier work. We are grateful to those colleagues who made possible an inter-institutional collaboration — Dame Margaret Weston, Director of the Science Museum, and Dr R.G.W. Anderson, Keeper of Chemistry, who gave support to this enterprise, and Professor C.A. Russell, Head of the Department of History of Science and Technology at the Open University. We must also thank the librarians and archivists who have so kindly put up with our enquiries for obscure works over many years. We owe a particular debt of gratitude to the libraries of the Open University, the Royal Society of Chemistry, the Science Museum, the University of London and the Wellcome Institute. Mrs Jeanne Pingree, College Archivist at Imperial College, and Mrs Janet Percival, Archivist at University College, were more than usually generous.

A draft of the text was rigorously scrutinised by Dr R.G.W. Anderson, Dr W.H. Brock, Mr M. Bud and Mr J.B. Morrell. For their advice we are grateful. For the errors that remain, we alone are responsible.

We are grateful to Dr W.H. Brock for permission to cite from the transcription and translation of the Liebig-Hofmann correspondence edited by himself and the late Dr E. Wangemann. We thank Elizabeth Bonython for permission to cite from the Cole papers. Transcripts of Crown-Copyright records in the Public Record Office appear by permission of the Controller of H.M. Stationary Office. For permission to cite from manuscripts in their keeping we thank: the Bayerische Staatsbibliothek, Munich; the British Library Board; the Trustees of the British Museum (Natural History); Cheshire Record Office; the Library, Edinburgh University; the Institution of Electrical Engineers; the Institute of Geological Sciences; Liverpool City Libraries; City of Manchester Cultural Committee; the National

Library of Wales; the Royal Society of Chemistry; the Senate, University of London; the University of Strathclyde; the Library, University College London; the Victoria and Albert Museum; and the Wellcome Trustees, London.

It is traditional in prefaces of this sort to make acknowledgements to spouses. We are now both acutely aware of the reality behind these. The moral and intellectual support of our own families is appreciated.

To our families

Introduction

In early 1980 a manifesto, signed by a hundred and forty-two leading citizens, announced a new initiative of the Royal Society of Arts.[1] Entitled 'Education for Capability', the manifesto expressed concern about the structure of British education. An entrenched division between education and training was condemned: 'Thus the idea of the "educated man" is that of a scholarly leisured individual who has been neither educated nor trained to exercise useful skill... It is significant that we have no word for the culture that the Germans describe as "Technik" or the mode of working that the French describe as a "Metier"'. This highlighted a view of the relationship between abstract knowledge and practical skill which the signatories saw as endemic to the British education system. That relationship was already the focus of intense discussion begun more than a decade before. Scholars and bureaucrats concerned with research policy were investigating the interplay between advances in abstract knowledge and in technological performance. They began to question the view that scientific research had an automatic pay-off in technical innovation.

Little attention has been given to the historical origins of such assumptions. It is to the improved understanding of this history that our book is directed. As a perspective on the problem we have chosen the development of one of the most 'practical' of sciences in one of the most pragmatic of nations: British chemistry in the mid-nineteenth century.

The question of the relation of science to practice is naturally of more than British concern. Throughout the world exponentially increasing costs of research during the 1960's focused concern on the exact utility of scientific research. While the resultant debates over research policy were not themselves concerned with the origins of the assumptions then being criticised, they did highlight the complex patterns of dependency between belief and institutional structure. The debates began in the United States as the world's leading sponsor of research. In 1966 a report to the Department of Defense suggested that innovation in weapons had depended only slightly on scientific 'events'.[2] The scientific community then felt under fire but later felt defended by a subsequent report commissioned by the National Science Foundation which demonstrated that, contrary to the previous study, scientific advances had been crucial to the advance of technology.[3] Analagous disagreements were aired when in 1971 the British government was the recipient of contrasting documents dealing with science policy. Two reports of separate policy committees, headed by Sir Frederick Dainton and Lord Rothschild, were published together. Starting from different premises about the nature of scientific and technological linkages, they reached quite different conclusions about desirable modes of research funding.[4] The Dainton Report, arguing for the primacy of autonomous research councils, based its conclusions on the integrity of science; any divisions between pure and applied were arbitrary and unhelpful.[5] Rothschild, by contrast, argued that while basic research was concerned with an increase in knowledge, useful research was directed to goals defined by others. 'The government should therefore resist the view that there is no logical division between pure and applied research, a view that may be intended to protect the Research Councils from the imaginary ravages of applied R. & D. users.'[6] This brought forth powerful complaints from the scientific community concerned about the institutional repercussions.[7]

The debates about policy brought in their wake scholarly discussion about the historical relationship between scientific research and industry. A variety of investigations tended to show connections that were less direct and more complex than hitherto assumed.[8] The historian Edwin Layton argued with particular reference to the USA that in fact science and technology were not so much an intellectual continuum as 'mirror-image twins'.[9] They were each separate communities whose interests might be similar but whose values were opposite: scientists valued theories, whereas engineers were 'doers'. Layton argued that though during the nineteenth century the previously disparate communities of technology and science had become 'coupled', the scientific style of the engineering profession as it developed during the late nineteenth century should not blind us to essential differences in attitudes.

Layton's arguments could apply with similar force to European countries by which the United States had been much influenced:

Germany, France and Britain. Even brief summaries of developments in these countries indicate broad international patterns, yet also demonstrate fundamentally national dynamics. In Germany the higher education system, traditionally dominated by the universities, was increasingly split after 1870.[10] A hitherto minor branch, the Technische Hochschulen, with practical goals grew quickly. Universities which had had an important role in professional training increasingly glorified the purity of the knowledge they produced. The split in attitude between universities and Hochschulen was illustrated by deep disagreement over the permission granted in 1900 to Hochschulen to grant doctoral degrees.

Meanwhile in France an analagous division emerged from a different background.[11] With an education system dominated by the Grandes Ecoles and traditions of professional education, it was the introduction of purely scientific institutions that represented novelty. As part of a wholesale reorganisation of the education system, the Ecole Pratique des Hautes Etudes was established in 1868 to stimulate research. In the United States the National Academy of Science was being made over to researchers at the same time. The Academy founded in 1863 was soon devoted to abstract science. Edison, though an American hero, was never elected. Nonetheless while American scientists were multiplying and increasingly able to follow purely scientific paths, they found it necessary to maintain a rhetoric of the utility of science.[12]

Developments in Britain occurred roughly synchronously. After 1870 the teaching of science was divided increasingly between 'pure' and 'applied'. Donald Cardwell's *Organisation of Science in England* identified the Samuelson Committee of 1868 and the Devonshire Commission of the early 1870s as together a turning point in scientific affairs.[13]

It cannot be doubted that such comparisons indicate underlying trends. But the effect of transnational influences has never been simple to understand. Even in direct transmission ideas and institutions are modified and often transformed by the receiving culture. A recent survey highlights the changes in nature of the academic system when transferred from Germany to America.[14] Moreover, far from being ready for a proper multi-national understanding, the emergence of the concepts of pure and applied science within individual cultures has yet to be clarified. Explicitly this book is concerned only with Britain. At the same time it is to be expected that those whose expertise and interests are focused on other countries will recognise parallels to processes with which they are familiar.

Choice of a single national context does not in itself ensure the homogeneity of the culture examined.[15] By the nineteenth century Britain had become a complex and variegated society. Even the word 'scientist' highlighted not the fact of a unitary scientific community but the diversity of the followers of science.[16] One way to delimit a coherent object is to study the history of a specific discipline. Characteristically nineteenth-century disciplines had their own culture, transcending particular institutional contexts. Consideration of them enables us to examine both intellectual and social change.[17] As a particular level of scientific organisation, the discipline is to be distinguished from two other levels: research areas and specialties. Research areas are the short-lived conjunctures of problems and investigation that are the immediate context of advancing knowledge. More enduring are specialties with their particular skills, traditions and often visionary leaders. They have considerable intellectual integrity. Not so disciplines — embracing varied specialties, their importance is institutional rather than primarily intellectual. Social and cultural considerations lead to the grouping of diverse participants under such banners as 'Chemistry'.[18]

In recent years there have been several calls for discipline history and a growing monograph literature focuses on disciplines in individual national contexts. Hufbauer has argued the growing disciplinary identity of chemistry in eighteenth-century Germany.[19] Kohler's recent study of biochemistry explicitly deals with the development of his subject as discipline.[20] Kevles's *The Physicists* portrayed the complex growth in America of the heterogenous community of practitioners and their concerns in what is arguably the twentieth century's most influential science.[21]

Some of the respect shown to physics in the twentieth century would in mid-nineteenth-century Britain have been more readily accorded to chemistry. 'Where science was spoken of,' averred Captain J. F. Donnelly, responsible for the expanding system of science classes in 1868, 'it was generally supposed to mean chemistry.'[22] Until 1867 it was the principal subject of Science and Art Department classes, and it was the favoured subject for London BScs throughout the rest of the century. In Britain alone the chemical discipline grew dramatically. Over the decade of the 1830s its scale, in terms of practitioners or papers, was still measured in tens. Growth to 1870, by contrast, whether indicated by elections to membership of the Chemical Society, or by the number of papers published by members, or by the number of chemical patents was two orders of magnitude greater.[23] Chemistry was to be measured in thousands. Having been a focus for popular interest in science at the beginning of the century, it became the major academic science in mid-Victorian Britain.

Chemistry is specially appropriate as a focus for the study of the categories of 'pure' and 'applied' science. Perhaps more than any other discipline, it was born out of the relationship between theory and practice and was continually expected to mediate between them. Hannaway has shown that chemistry emerged in Europe during the late sixteenth century as the expression of the essence of a variety of crafts.[24] In 1853, three hundred years after Libavius had articulated chemistry in this way, the English chemist Lyon Playfair lectured the founder of the Art Schools in similar tones:

In plainer language your colours as well as your glazes are chemical appliances requiring as much practical skill as knowledge, that men of abstract science are insufficient to improve them greatly, but practical men acquainted with Science would exercise a more beneficial influence on them.[25]

This study follows chemistry from the early 1800s to the end of the century. We have, however, concentrated on the period 1830 to 1870. During that period the discipline came to be defined in terms of academic criteria, despite its varied practical contexts. Sustained by the faith of industry and the professions in the utility of chemical knowledge, academics became ever more important within the diverse chemical community. Through their teaching, research and institutional influence they delineated the scope of the discipline. Leading academic chemists advocated a division of labour between academics who were to teach and to promote pure science, and practitioners who were to apply it. Industrialists and professionals, though sceptical of the direct utility of collegial knowledge, could concur with the belief that what had been learnt in the cloisters might be regarded as the basis for subsequent practical training. These concepts underlay the enduring chemical curricula established during the 1850s and 1860s. They attracted considerable financial resources to chemistry when educational concepts of pure and applied science proved complementary to the pressing needs of teacher training. A series of government enquiries in the 1860s and 1870s mapped out the nation's ambitions for higher education for the rest of the century. Science teachers for elementary classes and future managers would be trained together in science colleges in provincial centres inspired by a great central institution in South Kensington. They would be trained together in pure science. Those who needed to understand technology would learn to apply their science in their subsequent careers. The concepts of 'pure' and 'applied' science were therefore part of the Great Instauration of the late nineteenth century. They justified the institutions and in turn they acquired institutionalised validity. Ironically the discipline of chemistry, promoted as intrinsically useful, prospered as a 'pure science'.

Distinguishing rhetoric from reality is a generic problem for historians. The cogency of Victorian arguments and their resonance with modern convictions have often obscured the distinction. Our aim has been to alternate, as far as possible, between analysis of programmatic rhetoric and institutional realities. The first chapter serves to examine the roles of practitioners and the variety of early nineteenth-century attitudes to the relationship between science and practice. It suggests that during the 1830s a group of academics with a Scottish background and with a strong belief in the natural coincidence of interests of manufacturers and savants acquired national prominence. Chapter Two explores the implications of the rise of academics trained in Scotland and soon in Germany. During the 1840s academic, professional and practical interests came together in a variety of new institutions: most prominently in the Chemical Society, founded in 1841, and in the Royal College of Chemistry, established in 1845. Chemistry was promoted as the solution to practical problems and its success spurred the growth of the newly significant academic sector. However, as is shown in Chapter Three, syllabuses included technological topics only at the end of courses, while most pupils stayed only for the early elementary and fundamental lessons. The system was rationalised through new ideas about the proper relationship of scientific and practical knowledge. These ideas are then related to the disciplinary ambitions of the participants. The complexity of that relationship is explored in Chapter Four through prosopographical studies of Chemical Society members, of alumni of the Royal College of Chemistry, and of chemical patentees. In Chapter Five, attention moves back from chemistry to the politics of science education as a whole. The alliance between bureaucratic entrepreneurs, concerned to build institutions for science teachers, and academics seeking new support, led to the articulation and acceptance of the philosophy of pure and applied science. In Chapter Six we show how these prescriptions were carried out in the following decades. The chemical community became increasingly dominated by academics. Although a renewed concern about technical education led to more formal practical training, this was to be subsidiary to pure science. It was not that practical skills were in themselves disdained, rather it was generally agreed that such skills could not best be transmitted within the sheltered context of academe. By 1900 the encouragement of knowing rather than doing, so deplored by the 'Education for Capability' signatories, had become well established in the chemical curriculum of the British educational system.

Chapter One:
The background

The rhetoric of Francis Bacon has resounded for three hundred years: 'Now the true and lawful goal of the sciences is none other than this: that human life be endowed with new discoveries and powers.'[1] Even if the 'powers' of which Bacon wrote were originally intended to be spiritual rather than merely technological, they soon acquired a strictly utilitarian interpretation. The 'authority' of Bacon gave English science in particular a practical image. So by the time of the industrial revolution, science had been recognised as useful for over a hundred years.[2] When in 1802 Humphry Davy, the newly appointed Professor of Chemistry at the Royal Institution, gave his inaugural lecture he rhapsodised without fear of contradiction on the particular utility of knowing chemistry.[3] Nevertheless there was no generally agreed description of an exact relationship between science and practice in Britain at the time.

Not that the utility of science lacked interest. Nor were there well defined rival ideologies. However, the lack of an agreed intellectual framework reflected the absence of a central scientific type. In England despite its cultural and economic wealth there was no academic community of chemists. Though professors could be found at both Oxford and Cambridge they were poorly paid, part time, and often not in residence.[4] In London a university was only founded in 1826 and owed its ancestry to Scottish rather than indigenous models.[5] In Scotland there was a vital academic community but regional cultures also offered a variety of other possibilities for intellectual life. Hand in hand with distinct social structures went apposite assumptions.

It is conventional, if perhaps exaggeratedly schematic, to identify the scientific types with particular regions. The archetypical scientific cultures of London, of the North of England and of Glasgow can be delineated particularly clearly. In London, the great melting pot, quite different classes rubbed shoulders. Both a patriarchal elite rooted in the eighteenth century and a newer class of thrusting manufacturers believed that science should be useful.[6] Moreover faith in science offered opportunities to a new class of professional chemists whose livelihoods depended on persuading others of the utility of science.[7] A further possibility came to be particularly associated with the bursting new urban centres of the North of England. There, 'utility' meant the interrelated benefits reflected in moral worth and in economic success that were assumed to be inseparable from science.[8] In Scotland, where there was an established academic community with ideals of community service, science was conceived as a method whose full potential would be realised through an alliance of academics and manufacturers.[9] While these models were not themselves debated, the communities within which they were assumed jostled for influence. In this chapter we shall characterise the varied models of the early nineteenth century at greater length, leading on to the specific way Scottish academics acquired a new central position within the chemical communities of the whole island.

London

London in the early nineteenth century was both the largest city in the world and the capital of the nation. Its inhabitants included such disparate types of chemical enthusiast as manufacturers exploiting the novel possibilities of chemical techniques for profit and an old-established aristocratic elite. It is true that the two categories were not absolutely without overlap, nevertheless a wealth of social codes served to distinguish the true aristocrat from the merely rich. Membership of such clubs as the Athenaeum, in whose foundation Davy himself played a leading role, served to confirm distinction (as it still does).[10] Chemistry's two communities were effectively served by separate institutions. In view of the social division within the ranks of chemical devotees, the lack until 1841 of a specialist learned society, even at a time when sciences of comparable importance had become so served, is not surprising. A closer look at the origins of the early learned society shows that this lag did not simply mean that chemistry was 'late' to be recognised.

For the sciences generally, elite culture was expressed through the leading societies. These aimed to serve science and the nation rather than the individual. At the head of metropolitan science was the all-encompassing Royal Society. There were also an increasing number of specialised societies: the Linnean (f. 1788), the Geological (f. 1807) and the Astronomical (f. 1820) were perhaps the most important.[11] The characteristic functions of early nineteenth-century associations arose out of the new tasks that were set for science in the late eighteenth century – the systematic classification and collection of medical, living, mineral and astronomical components of the cosmos.[12] For example, the Linnean Society was established expressly for the cataloguing of fauna and flora according to the new Linnean classification. Furthermore it was considered that science should not be concerned solely with the classification of the world but also with its exploitation for the benefit of the nation.[13] Sir John Sinclair expressed the ideal well in 1811:

The labour of collecting and condensing useful information will soon be completed, so that he [his son] & his contemporaries will have nothing to do, but to apply the facts & axioms we have been collecting to objects of Public Utility.[14]

Such aims reflected the adaptation of traditional patriarchal attitudes to a world of rapidly changing technology and a developing economy. For instance, the Horticultural Society was founded in 1804 to foster the development of new species of plants, with ornamental and particularly with economic improvement in mind.[15]

The objectives that had brought about the Linnean and Horticultural Societies were combined in the establishment of the Geological Society in 1807.[16] It had the primary purpose of mapping the mineral wealth of Great Britain. The enterprise would bring resources to the notice of landowners, foster the discovery of new uses, and add to knowledge about the earth. It would be collective, entailing the labours of provincial observers as well as of metropolitan savants. Power struggles inside the Society soon meant that instead of attracting the support of geological afficionados from around the country, it became the preserve of a few metropolitan savants and many others of their own class – 'poets, statesmen, historians, and warriors'.[17] Most members were effectively patrons rather than practitioners. They supported the Society with their subscriptions and by their interests rather than through research. Despite a continuing rhetoric of utility, the mineralogical survey was abandoned and instead the interests of the Society turned to elaborate descriptions of particular areas.[18]

Therefore the rise of the learned societies cannot be seen simply as a consequence of the increasing complexity of individual specialties, or as a reflection of the emergence of fully fledged disciplines. Even where utility was not an immediate object, the new societies of the early nineteenth century were often established to further particular large projects that demanded either the participation or the patronage of large numbers of people.[19] In chemistry at this time the possibility of communal programmes of utility clashed with the search for personal advantage. Commercial interests constrained communication even within the practising community and at the same time they seemed alien to a patriarchal elite. On the other hand, within their separate communities individuals found many ways to deploy chemistry to advantage.

For the elite, as early as 1781, a fashionable chemical club visited by such guests to the metropolis as Benjamin Franklin is recorded as having met once a fortnight at the Chapter Coffee House.[20] The record of what was possibly a different chemical club can be traced from 1807 to 1826. As far as we know, no more than five men ever met at its bi-weekly meetings, and total traceable membership was only fifteen. In keeping with the title 'club', membership was exclusive. It comprised the two leading chemists in London, Humphry Davy and William Hyde Wollaston, also several leading physicians with chemical leanings and a few amateurs. Social acceptibility was clearly as important as scientific expertise. The well regarded chemists William Pepys and William Prout were refused admission, while men less well known for contributions to science, such as the politician Henry Warburton, a close friend of Wollaston and the landowner Sir John Sebright, became members. The club welcomed eminent guests such as Jacob Berzelius and John Dalton.[21] Davy developed his ideas on nitrogen trichloride after hearing news at the club about some experiments recently conducted. Not only chemistry was discussed: in 1816 members celebrated the repeal of income tax.[22] Such a small group of 'Philosophers, Chemists and Scientific Amateurs' was a congenial context to a man such as Sir Humphry Davy, President of the Royal Society from 1821 to 1827, who was happy to move in a small elite circle and had aristocratic pretensions.

By contrast the (Royal) Society of Arts made an attempt to harness the creativity of artisans. Established in 1754 to help the development of the arts and manufactures, the approach was characteristically paternal. Leading members constituted six committees, each with its own domain. Indicating its perceived utility, Chemistry was one of these, along with Polite Arts, Agriculture, Manufactures, Mechanics, and Colonies and Trade. Desirable inventions and discoveries were identified and solicited from the country's artisans. Solutions were publicised, and rewarded with medals and premiums of up to a hundred guineas. Among the problems set in 1820 were 'Refining copper from ore' and 'Preparation of sulphuric acid from sulphur, without the the use of any nitric acid'.[23] Probably because of the countervailing attractions of patents or commercial monopoly the society was in gradual decline during the early nineteenth century.[24]

The Royal Institution, founded in 1799, was also a paternalist institution that was intended to improve the agriculture and industry of the nation.[25] Unlike the Society of Arts, which drew on the genius of the nation as a whole, the Royal Institution would depend on the esoteric expertise of its own exclusive staff. Chemistry was allotted an important role because the subject was seen to be central to innovation in industry and in agriculture. A professor of chemistry was the first to be appointed, and he was assigned useful projects by the Committee of Management. Within a few months of his appointment in 1801 the young Humphry Davy was directed to acquaint himself with tanning and shortly thereafter with agriculture. Such down-to-earth activities were not incompatible with the advancement of more abstract knowledge, such as his discovery of new elements, and in no way prevented Davy from successfully becoming a member of the scientific elite.

However, the generation of useful knowledge for private gain was scorned. Davy's notebooks contain sardonic comments about men who were too obvious in their search for profit.[26] When the eminent chemist William Hyde Wollaston was revealed as the source of a hitherto unknown element, palladium, on sale at a high price as a curiosity, Joseph Banks, President of the Royal Society, wrote to a shocked friend:

The keeping of secrets among men of science is not the custom here; and those who enter into it cannot be considered as holding the same situation in the scientific world as those who are open and communicative; his reason for secrecy is however, a justifiable one; tho [sic] not such a one as either you or I should wish to avail ourselves of. He had purchased a large quantity of Platina, & wished to make money of it.[27]

By contrast, the other old-established class of chemists in London, the drug makers and small chemical manufacturers, saw chemistry as their expertise. Luke Howard, a leading chemical manufacturer of the early nineteenth century, applied himself assiduously to his hobby of meteorology but published nothing on chemistry. He explained himself in a letter:

It may appear singular that with such opportunities, I should have published nothing *as a Chemist*. The reply to such a proposal would be short, and decided. C'est notre *metier* – we have to *live* by the practice of Chemistry as an art, and not by exhibiting it as a science. The success of our endeavours, under the vigorous competition which every ingenious man has here, to sustain, depends on *using*, while we can do it exclusively, the few new facts that turn up in the routine of practice.[28]

Though Howard's success both as manufacturer and as savant marked him out, his attitude to chemistry was that of many tradesmen. It was antithetical to the public advancement and use of chemical knowledge favoured by men such as Davy.

The breadth of the social divide in metropolitan chemistry was illustrated by the history of unsuccessful attempts to found a popular chemical society. As early as 1806 Frederick Accum, a pioneer of gas manufacture and a popular lecturer who had assisted Davy for a time, announced plans for a large 'London Chemical Society'. Men with all levels of chemical competence were invited to join. Even the mere 'lover of science' could contribute valuable observations to men of 'higher acquirements'. In exchange he would learn the science of chemistry. The society would provide an audience for chemical ideas which could spur the otherwise isolated experimenter. Those with little time to peruse journals would be helped by prepared abstracts of chemical publications and correspondence. Despite, or perhaps because of, the ambition of the scheme, nothing further to the initial announcement has been found.[29]

The breach between the classes of chemical practitioners became explicit in arguments over the next attempt. In 1824 a new London Chemical Society was proposed to help practising chemists to study together. It was promoted by that ideologist of the skilled working man, Thomas Hodgskin, in his journal *The Chemist*. The society had a spectacular inauguration addressed by the father of the Mechanics' Institute movement, George Birkbeck, but expired shortly after.[30]

Hodgskin had sought the support of Davy for *The Chemist*, but was rejected out of hand. As Davy wrote to a friend:

The Chemist & Mechanics Magazine made overtures to me by sending their first numbers &c. The Chemist [illegible] exaggerated praises, but I scorn shake [sic] hands with chimney sweepers even when in their hay day clothes & when they call me 'Your Honour'.[31]

The bitter Hodgskin took the journalist's revenge and complained of Davy:

He stands aloof amidst dignitaries, nobles and philosophers, and apparently takes no concern in the improvement of those classes for whom our labours are intended and to whom we look for support.[32]

The clash of the two classes should not obscure the emergence of a third, with great long-term importance: the professional chemists. Broad surveys of the history of science have often emphasised a process of 'professionalisation' that could be seen in the nineteenth century.[33] It has been argued that the practice of science became the basis of a livelihood only as the number of academic posts grew. That approach employs a sociological concept of professions based on a collection of twentieth-century attributes.[34] This can confuse the student of nineteenth-century British chemistry. For the lack in England of an academic career, as we would recognise it today, did not mean that chemistry was therefore necessarily 'amateur'. After all, Davy through his professorship earned his living by chemistry. Particularly in London, opportunities emerged for gaining a livelihood through teaching and consultancy, and men who were sometimes styled 'professional chemists' took advantage of them. Though there was no organised profession of chemistry, chemists earned, sometimes very prosperous, livings through a combination of roles analogous to those of other professionals, such as lawyers and accountants. Until recently, few members of either of these professions either studied or taught in universities in Britain. Similarly, chemists were trained through apprenticeships and, having qualified, worked independently: consulting, earning considerable fees from their own apprentices, and often lecturing part time to candidates for examinations. These new 'professionals', on the whole neither of the elite nor manufacturers, straddled the two older communities.

The new role evolved as demands for chemical knowledge emerged during the early nineteenth century. Enterprising chemists found ways of turning occasional earnings into incomes. Though individually inadequate, earnings from a variety of sources could be combined to provide a comfortable life style. With the twentieth-century dominance of the academic model of teaching and disinterested research as the principal 'scientific' role, the professional career, so important in nineteenth-century England, has often been forgotten. Nevertheless, in the early nineteenth century, it sustained some of the most creative scientists of all time, including both Michael Faraday and John Dalton.

During the eighteenth century medical students had provided an established but small market for chemical lecturers. Aspiring apothecaries sought knowledge of chemistry not only for the light it might shed on physiology and pharmacy, but also for the gloss of learning it would bestow. During the first fifteen years of the nineteenth century, between four and six chemistry courses for medical students were annually available in London.[35] On the whole, the lecturers were physicians who also taught other subjects. This situation was transformed by the medical reform of 1815, which charged the Society of Apothecaries and the Royal College of Surgeons with licensing apothecaries and surgeons throughout England and Wales. The two bodies established regulations which demanded that students not only take examinations in chemistry but also enrol in lecture courses. Requirements escalated steadily. Before 1829 the Society of Apothecaries required one class in chemistry. By 1828 fourteen were offered annually in the metropolis, more than twice as many as before the Apothecaries Act of 1815.[36] From 1829 to 1835 the Society required two classes, and, from 1835, practical chemistry as well. The number of students multiplied, and consequently so did the number of chemistry courses. A survey of 1836 showed that there were 640 medical students in London and 350 in the rest of the country.[37] Michael Faraday and his colleague W. T. Brande, who taught medical students at the Royal Institution, had annual enrolments of between fifty and a hundred students. Satisfactory emoluments accrued to the teachers, who charged £4 to £7 a class. The joint lecturers at Guy's Hospital earned £670 from their chemistry courses in 1820.[38]

The men who emerged as leading chemistry teachers after the 1815 Act characteristically followed rather different careers from their predecessors. Firstly they specialised in chemistry and did not teach a wide range of subjects. This new practice was given powerful reinforcement by the Society of Apothecaries in 1830. Regulations were established for the certification of teachers.[39] No registered lecturer was to offer more than two subjects, applicants for certification were required to produce testimonials from two respected colleagues, and lecturers in chemistry were required to have apparatus and a laboratory. Secondly, the new teachers were not practising medical men, though most had been trained in medicine. This withdrawal from medicine reflected the new availability of other markets. Outside medicine, there were other extensive lecturing opportunities. At the top end of the market there was the Royal Institution (and lesser imitators). Occasional lecture series earned £50.[40] Beyond that was a host of organisations that sponsored chemical lectures. Within the metropolitan area alone, there were a dozen mechanics' institutes in 1839. By then, there were eighteen London-based lecturers who offered popular chemistry courses.[41]

In addition to teaching, there were new opportunities for consultancy. The range was indicated by J. T. Cooper, apothecary turned chemical manufacturer turned lecturer and consultant. In 1827 he applied for a Chair in Chemistry Applied to the Arts at the University of London and supplied a *curriculum vitae* in support of his application. In it he testified:

That my pursuits have been wholly directed to chemical science for the last twenty years — That I have been a public teacher of Scientific and practical Chemistry in London for fourteen years at various institutions, Societies and at other places — That having been continually resorted to by various individuals in a professional capacity, for opinions and advice — as relating to the taking out of Letters Patent — giving evidence before Committees of the House of Commons, and in Courts of Law — and by being consulted in confidence by manufacturers as to the improvement of their processes; and enjoying the friendship and confidence of numerous individuals engaged in various processes of chemical manufactures — by these various means I have been brought into more immediate connection wih the principles and practices of the following chemical arts ...[42]

There followed a long list of industries arranged under thirty main heads, including 'smelting and refining metals as practised on the large scale', 'the preparation of pigments' and 'the process of tanning'. The range of activities were quite typical for the metropolitan professional chemists. As one of the most successful, Faraday had an average consulting income of some £240 a year during the 1820s and '30s. At its height, in 1831, his consultancy brought him over a thousand pounds. In 1836, accepting an invitation from Trinity House to act as their advisor, he wrote: 'In consequence of the good will and confidence of all around me, I can at any moment convert my time into money.'[43] One did not have to be Faraday to prosper. Cooper lived well if not luxuriously. In 1841 his household of four was served by two servants. His neighbours were brewers, solicitors and civil servants.[44]

Faraday's status as a consultant indicates the significance of the professional career to the development of science in England. The examples might be multiplied. The propounder of the atomic theory, John Dalton, followed such a path in Manchester.[45] However, it must be admitted that the search for an income was not always conducive to scholarly investigations. W. T. Brande, one of the most successful professionals, failed to fulfill early promise as a scholar. He once complained to Faraday about the difficulty of keeping up with modern developments. His friend, the eminent surgeon Benjamin Brodie, remarked that Brande's limited stature as a chemist reflected the labours necessary to keep a wife and family in style.[46]

The range of opportunities exploited by the consultants was opened up by the rapidly changing society and economy. As Morris Berman has argued, in a 'rational society' science was the preferred mode of solving society's problems.[47] Moreover many industries faced technical problems for which there were no traditional solutions. Whatever our prescriptions, the early nineteenth-century professionals saw questions as diverse as accident prevention in coal mines and economy in gas manufacture as peculiarly chemical problems.

The gas industry displayed dramatically the rapidity of technological change during the industrial revolution, and the novel problems encountered. It illustrates the congruence of the interests of the industrialists, in search of rapid solutions to intractable problems, and of chemists in search of a living. The manufacture of coal gas began about 1800. At first it was scarcely more than an instrument of amusement, but in 1805 Boulton and Watt supplied the huge Manchester cotton mill of Phillips and Lee with a plant that produced 11 million cubic feet a year.[48] In 1812 the Chartered Gas Company began the public supply of gas in London and built works whose size was of a new order of magnitude. In 1822 the plant in Horseferry Road produced 111,384,000 cubic feet − ten times more than Phillips and Lee's unprecedented plant of 1805.[49] Scale continued to increase. Production of gas in the metropolis doubled between 1822 and 1827, and doubled again in the following decade. The sums of money involved were vast. By the late 1830s the twelve London gas companies had invested £2,800,000 in works, tanks, gasholders and apparatus.[50]

The process of gas manufacture entailed heating coal in a retort. The emitted gas contained impurities – bituminous vapour, ammoniacal liquor, essential oils and sulphuretted hydrogen. The gas was purified by condensation to remove volatile components and washed to remove oils, ammonia and some of the sulphuretted hydrogen. Finally the sulphuretted hydrogen was removed by passage through either lime water or dry lime. The gas was then stored in a gasholder for later distribution. The process and its evaluation were beset by problems of detail. It was not easy to measure the quality of gas produced. This was done by evaluating the candle power of a standard flame; but to ensure comparability it was also necessary to determine the specific gravity of the gas.[51] Again, coal was a major expense for a gas company and chemists, such as Brande at the Royal Institution, investigated the appropriateness of different kinds of coal for gas manufacture.[52] Another problem that caused the gas companies to come to the chemist was the disposal of by-products. An example is to be found in the records of the Chartered Gas Company. The foul water used to absorb ammonia liquor and hydrogen sulphide was disposed of by pouring it over hot coals. This was a convenient way of dousing the furnace but created a terrible stench. In April 1837 the firm's own Committee of Works decided that the process had to stop. Two chemists, J. T. Cooper and Thomas Everitt of Middlesex Hospital Medical School, were called in to advise on some other way of disposing of the stinking fluid.[53]

The problems inherent in the production of coal gas and the hegemony of existing companies stimulated entrepreneurs to find other analagous products. Whale oil was a traditional fuel for lighting, and on heating it also yielded a gas. In 1819 W. T. Brande, hitherto an exponent of coal gas, was contracted to investigate the production of oil gas. Brande's professional colleagues also benefited from the attempt when several were hired to defend the established method. To the coal gas companies, the endeavours of the oil gas promoters represented an intolerable economic threat. When in 1825 the new Westminster Oil-Gas Company applied for a charter, five chemists were found to testify that the new process would be inherently expensive and thus against the public interest. The testimony of J. T. Cooper was appreciated so highly that the directors of the company gave him a reference citing this case when he applied for a professorship.[54]

The role of chemists in the development of the gas industry points not only to the importance of chemical expertise in the solution of technical problems internal to industrial concerns, but also to the importance of chemical testimony in Parliament and in courts of law. Such testimony could be remunerative. In the case of Crossley v. Beverly (1829), concerning an important gas-making patent, the fees of the scientific witnesses during one day in court came to £111.[55] The possiblities of high remuneration sometimes induced chemists to find results that were convenient for their employers. A well known barrister taunted scientific witnesses as 'men ready to give advice and patronage in favour of either side of any question for a suitable fee'.[56]

The complexities and ambiguities of patent law offered great opportunities for the exploitation of chemical and legal expertise. Although taking out a patent was an expensive and complex legal act, the number of patents grew rapidly during the early nineteenth century. Patent laws were clarified, and the rights of patentees increased. Drawing up a patent was a fine art, for there were several grounds on which it could be contested should the patentee sue for infringement. A few chemists, including Arthur Aikin, Secretary of the Society of Arts, acquired a wide reputation in the art of drafting.[57]

These metropolitan professionals had close links with both the social elite and the practising manufacturers, but their commercial competition did not make for harmony. Faraday composed a humorous poem about the failure of the professionals, for just this reason, to establish their own club in 1824. Chemistry would have to be declared out of bounds at their meetings: 'Henceforth then ye chemists all union suspect/ unless Bacchus ye court and Hermes reject'.[58]

The North of England

London was the centre of professional chemical activity in England. Outside London, before the 1840s, it was generally necessary to combine consultancy and the teaching of chemistry with a further career. For example Thomas Barker first lecturer in chemistry at the York School of Medicine, Curator of the Yorkshire Philosophical Society Chemistry Laboratory, and Consultant Chemist to the Yorkshire Agricultural Society continued his profession of surgeon.[59] Alternatively one could combine consulting and teaching with manufacturing, as did Thomas Richardson of Newcastle.[60] John Dalton of Manchester was unusual in living on just teaching and consultancy.

The professionals played several cohesive roles in the service of a chemical community. First, through their popular lectures on chemistry, they showed manufacturers that what they were 'really' engaged in was chemistry on the large scale. Between 1816 and 1850 up to nine books a year were published on chemical topics.[61] Among the most popular were Parkinson's *Chemical Pocket Book*, Henry's *Epitome of Chemistry* and Parkes's *Chemical Catechism*. There are stories, not all apocryphal, of how young manufacturers were converted to chemistry by the reading of a single influential text. Young John Mercer from Blackburn bought a second-hand copy of Parkinson's *Chemical Pocket Book* on the day of his wedding. Thereafter he was an enthusiast for chemistry and he made several major technological advances related to calico printing.[62]

If professionals attracted their manufacturing neighbours' attention through their writings, they also did so through personal connections. Personal influence is a far more potent persuader than mass marketing. In Sheffield a local chemical society was begun in 1829.[63] In Manchester, during the 1840s, James Young, employed by Muspratt the soda maker but with wider consulting interests, founded a Manchester chemical society.[64] However, the consultants were generally more concerned to spread an appreciation of chemistry as a powerful tool than to disseminate freely their own esoteric knowledge.

A new urban industrial culture of consumers of science developed particularly in the North of England, the centre of the new industrialisation. Manchester typified the culture and institutional structure common thoughout the North. Not only was it the exemplary site of the industrial revolution, it was also the centre of an entire region, the economic and cultural entrepot for a large number of small villages and towns scattered around the base of the Pennines. Towns such as Hyde, Ashton under Lyne, Glossop,

Bolton, Salford, Rochdale, Blackley, Oldham and Styal looked to Manchester as their cultural centre.

In Manchester, as in many northern towns by the 1820s, scientific discourse was centralised in the 'Lit. and Phil.' founded in 1781. Arnold Thackray has described the vision of its founders. Science had possibilities as: 'polite knowledge, rational entertainment, theological instruction, professional occupation, technological agent, value transcendent pursuit, and as intellectual ratifier of a new world order'.[65] Though the founders had been largely physicians and lawyers, the Society became increasingly a forum for manufacturers. By 1830, 90% of the thirty-three Mancunian manufacturing chemists were members.[66]

While the 'Lit. and Phil.' was dedicated to the presentation of original papers, it was complemented by the Royal Manchester Institution (f. 1823), in which lectures on various aspects of science were given to the upper classes and by the Mechanics' Institution (f. 1824), whose lectures served the lower middle class.[67] While chemistry was an important concern, it was poorly differentiated from other sciences. For several decades the symbolic leader of Manchester science was a chemist, the internationally famous John Dalton. Though best remembered today as 'father' of the atomic theory, Dalton spent much of his time seriously assembling meteorological information and teaching mathematics.[68]

If chemistry did have a special status this was because of its apparent relevance to industry. In the early nineteenth century a very broad range of northern industries were arguably illuminated by chemistry. The new national economy, transport facilities, and the emergence of a market with a new order of size in the textile industries, all meant that even old processes were carried on increasingly on a much larger scale than ever before. The soap industry was represented in 1785 by 971 companies, producing, on average, 16 tons each; by 1830 production had increased threefold but the number of companies had dwindled to 309. The calico printing industry printed 480 million yards in 1840.[69] There were also completely new industries, such as the bleaching powder and Leblanc soda industries, based on chemical transformations. New connections were made between these industries and the old. The soda industry supplied one of the critical raw materials for the soap and glass industries. For the strictest commercial reasons chemistry was seen as useful knowledge. It is true that the utility of chemical science to many of the problems faced by the operative chemist or the larger-scale manufacturer was often limited. Large vats, rusty pipes and unskilled workmen were very different from the glassware of the laboratory. That said, and it is a reality of which the historian needs constantly to be reminded, there was an increasing place for chemistry.

The experience of Hugh Lee Pattinson, of Alston in Cumberland, exemplified the variety of factors behind a North Country manufacturer's interest in chemistry.[70] He had been apprenticed as a soap maker to a leading Tyneside firm. Perhaps because of this background, though he was working as a mine manager at the time, he was consulted in 1828 by Dr J. B. Anderson, an Edinburgh manufacturer, on the possibility of entering the soap business. Pattinson replied that the price of soap, like that of all other commodities, was weak, but that science might save the day.

There is not one soap manufacturer in a thousand knows any thing at all about his trade scientifically. That is of the action of alkalies upon fat and this is of the utmost consequence for conducting the process with accuracy and success so as to produce the best article at the cheapest rate.[71]

Two months later Pattinson was even more enthusiastic about the commercial advantages of chemical expertise. He had been invited by Anderson to join a venture to make sulphuric acid, alkali and soap in Bonnington, near Edinburgh:

Now the soap trade can be carried on to a great extent almost to any extent and if we were prosperous there is no saying what it might become. Besides this both of us having a knowledge of chemistry much superior to most manufacturers other branches might be added in time which we do not now contemplate.[72]

Pattinson's claim to a knowledge of chemistry was no mere boast. Some years earlier he had translated, apparently for his own use only, the papers on the nature of fats by the French professor, Chevreul. To do that, he had had first to teach himself French. His notebooks indicate elaborate experiments to perfect the manufacture of soap.[73] Five years later, while managing a Newcastle lead refinery, he developed a method of separating lead from silver, thereby solving a key industrial problem of the time. This proved a most lucrative discovery: he is reputed to have made £16,000 from it.[74] Later he went back to soap production and then into alkalis. His

Felling chemical works became one of the largest on Tyneside. So, for Pattinson, his knowledge of chemistry was of considerable financial advantage. And money mattered to him. He wrote to Anderson that he was a lover of £500 a year.[75] It would be obvious enough to suggest that his interest in chemistry had derived from his commercial concerns. However, it was more general.

Pattinson had been brought up in the small and remote town of Alston as the son of a successful Quaker grocer. From an early age, he is reputed to have taken an interest in science. There survives from 1819, when he was twenty-three and still in Alston, the manuscript of two lectures on chemistry that he delivered at a small village nearby. It gives a remarkably vivid picture of the character of chemistry as moral entertainment, cosmic vision and useful knowledge: qualities that were intimately connected. His account of affinity included the euphoric verse:

> The very law that moulds a tear
> And bids it trickle from its source
> That law preserves the earth a sphere
> And guides the planets in their course.[76]

These sentiments were followed by an exposition of the utility of chemistry in the analysis of iron and in zinc smelting.

Moving to Newcastle a few years later, and having taken up management of a soap works, Pattinson became an active member of the Newcastle Literary and Philosophical Society. With his friends he conversed, as he remembered when back in dreary Alston, on the 'poetry of Byron, the eloquence of Brougham, the philanthropy of Owen, and the science of Davy'.[77] In 1838 he would be vice-president of the Chemical Section of the British Association at its Newcastle meeting. Later in life he became fascinated by astronomy and built a telescope. Only in the scale of his achievement was Pattinson unusual.

Interested as he was in chemistry, and indeed in lecturing on it, Pattinson's commitment to such activities was strictly constrained by the demands of his manufacturing activities. In 1838 he was invited to become lecturer in chemistry at the new Newcastle Medical School. Although the job would have been only part-time, he felt unable to give his time to such work and courteously refused the offer.[78]

Scotland

The only opportunity for men such as Pattinson to be other than
self-trained in the science of chemistry was to study in Scotland.
There, the five long-established universities provided an education
to manufacturers as well as to the social elite. Dedicated to both
professional training and to general education, professors in
Scotland played central roles in the scientific cultures of their cities.[79]
Many of the pupils of the Scottish universities were medical
students. Nevertheless in Glasgow there had been since the middle
of the eighteenth century professors of chemistry who did not see
their clientele as restricted to medicine. William Cullen, lecturer in
chemistry at Glasgow from 1747 to 1756 and thereafter professor
first of chemistry and then of medicine at Edinburgh, emphasised
the general utility of chemistry. Cullen's message was handed on by
his disciple and successor at both Glasgow and Edinburgh, Joseph
Black.[80] In the introduction to his lectures, posthumously published
at the turn of the nineteenth century, Black clarified his notion of
chemistry. He was anxious to differentiate between scientific and
trading chemists. The latter carried out their manufacturing
processes by tradition. He would not lump them together with men
such as Lavoisier. On the other hand, though he did not 'apply'
scientific theory, Josiah Wedgwood was to be classified as a
'scientific chemist'. The crucial factor was method.

In like manner, we find numerous operators who, either with their own, or
by the hands of others whom they employ, exercise the various branches of
the valuable art of pottery. These persons, by an apprenticeship, or
otherwise, have learned to choose and to mix the proper materials; how to
form the vessels; to apply the glazing and other decorations; and lastly, how
to give the proper degree of fire to consolidate and finish the ware. These are
all artists, while they only exert in practice the skill they have acquired,
whether by communication from others, or by efforts of their own
ingenuity. But if there be a Wedgewood [sic] among them, who takes
pleasure in attaining more extensive knowledge of the subject, who by
making new trials, and varying the composition, the glazing, the firing, and
other parts of the process, endeavours to make improvements upon the art,
or to understand it better than before, such a person, in my opinion, is a
philosopher, or a man of thought, study, and invention.[81]

In other words to be scientific was to experiment, to reason and to
use what knowledge was available in the area. It did not necessarily
entail the application of esoteric theory.

A similar willingness to see industry as a legitimate context for the practice of science was to be seen in the long-lasting Thomas Thomson, Professor of Chemistry at Glasgow from 1818 to 1852.[82] Thomson promulgated a close relationship between his powerful academic centre and the technological environment of Glasgow. Manufacturers, already quite successful, such as Charles Macintosh and Walter Crum came to the university to take classes, while many younger students went out into industry. Outside the university, the Glasgow Philosophical Society offered another opportunity for academics and local industrialists to meet.[83] There was too the Andersonian Institution, in which technical classes in chemistry were given by Thomson's rival, Andrew Ure, eulogist of the factory system.[84] In Glasgow the social reality of an academic centre with close relations to surrounding industry had been reflected in a suitably pragmatic notion of science which encompassed both theory and method.[85] In the English industrial areas, the technical utility of science was subsumed under the more general moral utility. The special contribution of chemists in the English metropolis meanwhile had been the identification of the variety of customers for science and the means of serving them.

During the 1830s a small coterie of academics associated with Glasgow moved to the centre of British chemistry. They brought with them their vision of the relationship between science and practice. Less exclusive than those traditionally dominant in London, that division would make possible a new coalition of interests including manufacturers, professionals and academics.

Chemical debates

The Scottish ascendency emerged through vehement debates about the content of chemical science. Glasgow, under Thomson's influence, promoted the European programmes associated with Berzelius, the unchallenged doyen of chemistry in the Continental centres during the 1820s and '30s. In London meanwhile, eminent savants such as Faraday held very different views on a range of central issues. In order to understand the growth of influence of the Scots it is necessary to elucidate the chemical issues at stake as well as the novel institutions through which they were discussed.

Berzelius had developed a powerful model of chemical structure.[86] He had integrated major discoveries of the chemical revolution:

Lavoisier's oxygen theory of acids, an electrical theory of chemical affinity and Dalton's atomic theory. According to Berzelius, the atom of each element was electrically charged. All atoms, except those of oxygen, which had only a negative charge, had both positive and negative charges in unequal amounts. When an atom or group of atoms with a net negative charge was combined with oxygen, the result was an acid; if the net charge was positive the result was a base. This theory was developed in conjunction with a research programme concerned with the systematic analysis of minerals. In the early 1830s it was also, by analogy, applied to the products of the animal and vegetable world, initially by Liebig and Wöhler. Berzelius promulgated his theoretical views of chemical structure through a new quasi-algebraic symbolic system. Like the theory it was widely adopted on the Continent from the late 1820s.[87]

The systematic attempt by Berzelius to reduce mineralogy to chemistry had only one parallel in Britain. That was the enterprise of Thomas Thomson.[88] In an attempt to substantiate his own system of chemistry, Thomson faced experimental problems and issues similar to those of Berzelius. He was an early proponent of Dalton's atomic theory and then adopted Prout's theory that each elemental atom was composed of a characteristic number of fundamental identical atoms. These combined to give atoms of elements which, in turn, combined to give 'atoms' of salts. Thomson attempted to substantiate his theory by measuring the weight of atoms at various levels of complexity. He published his findings in 1825 in his *An Attempt to Establish the First Principles of Chemistry by Experiment*.[89] In practice, Thomson's programme entailed investigations similar to those of Berzelius — the analysis of salts and the measurement of atomic weights. At the same time, Thomson tended to pay much less attention than was traditional in English chemistry to matters of electricity. Not only philosophically based, his approach, as Thomson pointed out, reflected the bureaucratic organisation of his university. Electricity at Glasgow was the responsibility of the Professor of Natural Philosophy. He believed that questions of heat, light, electricity and magnetism should be brought together, as they were in France, under the rubric 'physique'.[90]

Thomson's interests were particularly important because of their influence on his pupils. As early as 1817 Thomson wrote to the Edinburgh Professor of Natural History, Robert Jameson, of his intention 'to establish a real chemical school in Glasgow and to breed up a set of young practical chemists'.[91] In addition to those graduates who became manufacturers, a few students became academic chemists who, in pursuing science, earned their daily bread principally through teaching, and a high reputation through research. The orientation he gave his students prepared several for a sympathy with the aims, methods and nomenclature of Continental chemists.

The most active authors among Thomson's pupils were three men who had studied during the 1820s − James Finlay Weir Johnston, Thomas Graham and Robert Dundas Thomson, a nephew of the teacher. During the 1830s Johnston published forty papers, Graham published seventeen, and even the youngest, R. D. Thomson published eighteen.[92] The productivity of these men was eased by academic positions and incomes that allowed them time to spend on unremunerative research. They found new posts. Johnston was fortunate to make a wealthy marriage which supplemented his meagre salary as Reader in Chemistry at the University of Durham, founded in 1833. Graham was professor first at the Andersonian Institution in Glasgow and then, from 1837, held the lucrative professorship of chemistry at University College London founded in 1826. In the latter position his income from student fees was more than £400 a year − quite substantial for a bachelor in his position.[93] During the later 1830s R. D. Thomson was a physician in London and teacher at the Blenheim School of Medicine. He returned to Glasgow in 1841 as an assistant to his uncle.

Though all three did study elsewhere after leaving Thomson's nursery, they continued to follow their early orientations. The two older men, Johnston and Graham, found teachers who were also concerned to assimilate mineralogy to inorganic chemistry. Johnston spent time in the laboratory of Berzelius in 1829, and in 1832 Graham moved to Edinburgh and studied with Edward Turner, a pupil of Stromeyer, a leading German mineralogical chemist. R. D. Thomson edited his uncle's journal, *Records of General Science*, during the early 1830s, was introduced to Liebig in 1837 by his uncle, and in 1841 studied in Liebig's laboratory. Similarities between programmes were complemented by mutual interests in certain intellectual tools. The concept of the 'atom' was acceptable to both schools. And in 1831 Graham was among the first in Britain to use Berzelius's symbolism.[94]

By contrast to Scottish pragmatism, according to the dominant tradition of metropolitan science, chemistry was the source of knowledge about the essential nature of matter. That tradition was upheld by Humphry Davy.[95] His powerful personality reinforced the traditional loyalties of lesser men. As late as 1837 the well known professional chemist Richard Phillips concluded a lengthy survey of chemistry with an extensive discussion of Davy's work, and only a miscellany of facts discovered by other men more recently.[96] Davy's authority and influence were increased yet more by the reputation of his pupil, Michael Faraday. In the *History of the Inductive Sciences*, published in 1837 by William Whewell, the period from 1810 to the late 1830s was dealt with in a chapter entitled 'Epoch of Davy and Faraday'.[97]

The grand system of Berzelius entailed claims about a level of chemical structure that was not susceptible to direct observation. To Davy, Faraday and their followers, the image of solid polar atoms bound by electrical force was probably wrong and certainly unsubstantiated.[98] That electricity could decompose chemical compounds meant merely that electricity was, on the gross scale, the analogue of chemical affinity on the micro-scale. The two forces might have a common cause, but they were not identical. Berzelius might discuss atoms, but in London Dalton was revered not for atomic theory but for his theory of chemical proportions. On the few occasions when Davy and Faraday allowed themselves to speculate publicly on the nature of the micro-level of chemical phenomena, they disclosed an interest in the force-based point atom theory of Boscovitch.[99]

Differences in theory were associated with divergence in programmes. No metropolitan chemist emulated the systematic analysis of minerals pursued by Berzelius. The connection made by the Swede between chemical affinity and mineral composition could not be followed by chemists who eschewed the Berzelian theory of chemical affinity. Instead in London attention was drawn to the phenomena of electricity, the observable analogue of affinity. Massive batteries were constructed and the major advances ranged from Davy's dissociation of alkali salts to Faraday's discovery of the principle of the electric motor.

The system of formulae adopted by Berzelius was opposed until the 1840s. A main antagonist was William Whewell. He criticised the Berzelian symbols thus:

Its formulae are merely unconnected records of inferences which are in some degree arbitrary; the analysis itself, the fundamental and certain fact from which inferences are made, is not recorded in the symbol; and the connexion between different formulae, the identity of which is a necessary and important circumstance, can be recognised only by an entire perversion of all algebraical rules.[100]

For Whewell, formulae were to be seen as shorthand statements of analytical results rather than as statements of chemical structure. Accordingly he worked out a rigorously algebraical system. Faraday complimented Whewell on his contribution and suggested that the Berzelian nomenclature was but a thinly veiled attempt to promulgate the complementary theoretical system. No doubt there was at the back of his mind the precedent of Lavoisier, who had promoted the new theory of elements through a novel nomenclature. Another factor, raised by Faraday, was the esoteric nature of the formulae; they were not accessible to the layman.[101] The objection had obvious significance for men who saw the audience for chemistry as the educated gentleman. Anyone who understood the principles of algebra could follow Whewell's symbolism, whilst Berzelius's symbols were only for the chemically initiated. Faraday's colleague, W. T. Brande, incorporated Whewell's formulae into the fourth edition of his *Manual of Chemistry* (1836) and J. F. Daniell used them in his *Introduction to the Study of Chemical Philosophy* (1839), which was dedicated to Faraday.

The difference between the chemistries promoted by leading protagonists in the metropolitan and Glasgow schools may be highlighted by contrasting the works of two chemists. The text by Daniell, Professor of Chemistry at King's College London, was entitled *Introduction to the Study of Chemical Philosophy*. Published in 1839, it indicated how chemistry ought to be studied from the London perspective. The atomic theory was mentioned only at the end as 'best kept out of sight in the first steps ... as likely to turn the mind from that rigid method of induction from facts, by which alone the student can be safely guided'.[102] Where formulae were used they were of the Whewellian type. The first 265 pages were devoted to what we would perhaps call physical phenomena – force, light and electricity. Then there were only 120 pages on chemical affinity and the characteristics of chemical compounds, followed by a further 120 pages on electricity. Contrast the text by Thomson's pupil, Thomas Graham, *Elements of Chemistry: Including the Application of the Science to the Arts* which was published in parts between 1837 and 1842.[103] Graham devoted sixty-one pages to atomic theory and associated concepts – isomorphism, isomerism and atomic specific heats. There were only fifty-six pages on electricity. Altogether in *Elements of Chemistry* one finds 149 pages on physical phenomena and 455 pages on the preparation and properties of inorganic compounds.

The reputation of Humphry Davy seems to have been sufficient to stave off criticism of London chemistry during his lifetime. However, a storm broke shortly after his death in 1829. The so-called 'decline of science' debate was largely focused on chemistry. Certain aspects revolved implicitly around the disagreement between metropolitan and Scottish chemists over the desirable direction to be taken by their science.

The first stone was cast by Thomas Thomson. In 1829 he published an article in the *Edinburgh Review*, 'History and Present State of Chemical Science', which sharply attacked British chemistry:

About twenty-five years ago, at least thirty individuals might have been reckoned in Great Britian actively employed in chemical investigations; now we can scarcely reckon ten. Some cause must exist for this retrogradation, so different from what is exhibited on the Continent, especially in France and Germany.[104]

Thomson attributed the 'retrogradation' to the weakness of practical training. Of course Thomson's own course was the exception to the dismal rule.

A year later, more specific attacks were launched. Two related lines of criticism emerged in articles published for the *Encyclopaedia Metropolitana*: English chemists were slow to participate in the Berzelian programme, and were excessively concerned with electricity. These were criticisms of the dominant metropolitan tradition, though they were put forward as a critique of insularity and backwardness. Francis Lunn, a Cambridge Fellow with mineralogical interests, concluded the historical portion of his article on 'Chemistry' with the observation: 'that during this period good Chemical analyses and researches have, with few exceptions, been rare in England.' He lamented that British chemists appeared to be concentrating excessively on electrical phenomena. On the other hand they paid insufficient attention to the theories and research programme of Berzelius. He asked: 'Are we sure we understand the nature of Nitrogen?'[105] (Nitrogen, an element in England, was considered an oxide of the Ammonium radical, NH⁴, on the Continent.)

The article by Lunn was followed in the *Encyclopaedia Metropolitana* by a contribution from John Herschel dealing with sound. Herschel had detested Davy personally and disagreed with his followers. Models, such as the atomic theory, were to be used according to their utility rather than only when proved 'true'. Long ago, he had been influential in the introduction to Britain of the heuristically powerful Leibnizian calculus notation. In a footnote to his article on sound he assaulted the provincialism of chemists in England:

Here, whole branches of continental discovery are unstudied, and indeed almost unknown even by name. It is vain to conceal the melancholy truth. We are fast dropping behind. In Mathematics we have long since drawn the rein and given over the hopeless race. In Chemistry the case is not much better. Who can tell us anthing of the Sulpho-salts? Who will explain to us the laws of isomorphism? Nay, who among us has even verified Thenard's experiments on the oxygenated acids — Oersted's and Berzelius's on the radicals of the Earths — Balard's and Serrulas's on the combinations of Brome — and a hundred other splendid trains of research in that fascinating science?[106]

Condemnation of British science reached a new intensity with *Reflections on the Decline of Science in England*, published in 1830 by Herschel's friend, Charles Babbage. The work cited the complaints of Herschel and Lunn, though it was mainly directed against the meagre rewards available to the British savant.[107] Herschel was also supported by David Brewster, the Scottish natural philosopher. In an anonymous review of Babbage's book, published in the *Quarterly Review*, Brewster quoted Herschel's derogatory comments at length and enthusiastically endorsed their sentiment. He gave wider currency to his article by giving it an adulatory review in another lengthy piece in the *Edinburgh Journal of Science*.[108]

Not surprisingly, the attacks were vigorously contested. Faraday published a letter, by the Dutchman Moll, that reflected on the greatness of English savants and on their honourable social status. Moll reminded his readers of the contribution made by Englishmen to knowledge of electricity, an area neglected by Berzelius.[109] Brewster in turn replied to Moll's argument in the *Edinburgh Journal of Science*.

In his inaugural address as professor at King's College in 1831, J. F. Daniell undertook a counter-attack against the 'declinists'. [110] He argued that the principal charges of inattention to sulpho-salts and isomorphism were misplaced. Both were rather dubious concepts. Daniell did not deny that English chemists had placed unusual emphasis on electrical research, he merely scorned the censure of such important work.

This skirmish over the state of British chemistry was the prelude to a more formalised and polite debate conducted in the 1830s in the novel forum of the British Association for the Advancement of Science, founded in 1831.

The British Association

The BAAS, by virtue of its 'Section B' devoted to chemistry and related sciences, was the first organisation to bring together chemists from throughout the country for specialised discourse. [111] Meetings attended by about a thousand people (and by about 2,000 at Edinburgh in 1834 and at Newcastle in 1838) were held annually in provincial centres. Covered fully by the press, local and national, they were media events. The visitors were so disparate that Whewell found it necessary to describe them by a new word, 'scientist'. Eight or nine sections (the exact number varied from year to year) were each responsible for a particular science, including chemistry, and each was overseen by an annually appointed committee. These committees were no mere administrative convenience, instead they were each to decide on the desirable directions to be taken by a particular science, and then to co-ordinate the activities of the entire scientific community for advance in the way required. Individuals were to be joined in an organised 'factory' of science. Overseeing the proceedings and giving order to what might have been chaos were committees consisting of a president, two or three vice-presidents, a similar number of secretaries, and a dozen or so others without office. Experts were deputed to report on the progress of worthy specialties and particular research projects were encouraged — sometimes with cash. Additional unsolicited papers were read and discussed. The importance and prominence of sponsored research and reports gave the specialised committees considerable influence.

On the broad stage, the British Association gave strategic influence to the coterie of its liberal Anglican leaders. The

'Gentlemen of Science' became national figures.[112] On a smaller scale, the Chemistry Section also bestowed a certain glamour and patronage on its elite. The section was dominated by eight men who sat on its committee year after year. They shared an adherence to atomic doctrines. Prominent among them was the cohesive school of Thomson, including Thomson himself, J. F. W. Johnston, R. D. Thomson and Thomas Graham. Notably missing from the octet were the traditional leaders of metropolitan chemistry. Faraday and Daniell only attended occasionally.

The leadership of the Thomson school encouraged strong links between Section B and Continental chemistry. Foreign visitors were welcomed to the Section and, from 1837, were even admitted to the committee. At the Glasgow meeting in 1840, five out of the thirteen members of a committee dominated by the Glasgow school were from the Continent. Similarly, foreigners came to play a considerable part in the reading of papers. The influence of Thomson and his pupils can be clearly seen in the direction of the proceedings. The solicited reports on the state and progress of chemistry are of particular interest. They were means through which the committees exerted their power and attempted to influence the orientation of chemistry in Britain. Right at the beginning, the authority of the metropolitan tradition was questioned. At the first meeting, J. F. W. Johnston was deputed to prepare 'a view of chemical science, particularly in foreign countries'.[113] In 1832 the committee requested information on those areas of Berzelian science that Herschel had singled out as ignored by English chemists in his *Encyclopaedia Metropolitana* article. Johnston was asked to present evidence for the theory of sulphur salts, while a group was assigned to report on isomorphism.[114]

In 1834 the problem of the divergence between metropolitan and European chemistry was faced again. Held in Edinburgh, it was the first meeting at which Thomas Thomson, R. D. Thomson and Graham were on the chemistry committee. The question of symbolism was discussed. As we have seen, the issue of which system to use was not merely one of convention, for fundamental theoretical considerations were at stake. However, in a clever manoeuvre, the issue was brought down to the level of practicality. In recent literature, ten different symbolic systems could be found. There was general assent to the proposition that 'Unless the evils resulting from every person who deems that he has made an improvement introducing new-fangled combinations be remedied ... chemical science would soon become a perfect chaos.'[115] A committee was formed to consider the question. The short report read the following year represented a clear victory for the Berzelian system, with only cosmetic deference to the opinions of Whewell and none at all to those of Dalton. The committee recommended that the system 'which is in general use on the continent' be adopted.[116]

In the Chemical Section of the British Association the Thomson School came to enjoy intellectual and social hegemony. Men such as Thomson himself were comfortable with the heterogenous audiences of the British Association and with the style they demanded. In 1839 Graham, while addressing the Chemistry Section, seemed to find no difficulty in describing the two major advances of the year as the new radical theory of Dumas and Liebig, and Gossage's new method of removing sulphur from waste emitted by soda works.[117] The following year Thomson welcomed visitors to Glasgow with a description of the area's chemical industries. Not only were they speaking from the country's only national pulpit of chemistry, but the Scottish academics were also expressing a catholic view of their science that could appeal to chemistry's several communities. The appeal of this view was shortly to be reflected in the success of the new Chemical Society, established in 1841, with Graham as the first president.

Chapter Two:
The emergence of an academic community

The transformation of British chemistry in the 1840s has been traditionally ascribed to the influence of the German chemist, Justus Liebig, and his British pupils. In Germany chemistry had become a dynamic and mature discipline within the universities as early as the latter half of the eighteenth century.[1] In the 1830s it grew with renewed vigour. Symbolic was the success of the young Justus Liebig, Professor of Chemistry at the small university of Giessen from 1825 to 1852. He was an able chemist who perfected a new method of organic analysis that was both accurate and much faster than previous methods. With this method, Liebig could train large numbers of students in analytical techniques. He was a persuasive propagandist too, arguing both in Germany and abroad for the multiple utilities of chemistry. The aggressive and charismatic Liebig came to symbolise the powerful leader to generations of chemists.[2] His bust still commands the entrance to the library of the Royal Society of Chemistry.

Liebig's laboratory attracted many foreign students, including Englishmen. A pilgrimage to Giessen, and at least a short spell of work in the laboratory there, became a part of a good chemical education in the early 1840s.[3] Through his writings and pupils Liebig acquired a high reputation in Britain. Nevertheless while Liebig's own influence and that of his pupils was certainly important there, the impact in itself of the German example on the evolution of British chemistry should not be overemphasised. The nascent academic community of British chemistry was strongly reinforced in the 1840s through indigenous institutional developments. The effect was not merely to multiply the number of chemists, but also to create a new category of chemist. No longer was the pursuit of esoteric chemical theories the province of gentlemen seeking polite knowledge; rather, it became a career in itself.

Academics became the chief advancers of the discipline, but both for social and for cognitive reasons they could not do without the

professionals and manufacturers. Alone they could neither provide a
forum for discussing research output, nor generate a sufficiently
large student body to justify academic posts. Nevertheless, the
academics did successfully constitute themselves the elite of a
chemical community, formed out of the more established practising
sector, by defining their area of expertise as research. This skill was
to be considered primary. With this new hierarchical structure even
the traditional professional community in London concurred. Its
leaders saw an association with a separate academic science as a way
of enhancing the status of the practice of chemistry.

During the early nineteenth century the diverse uses of chemistry
and the variety of practitioners had resulted in institutional
fragmentation. New umbrella organisations founded in the 1840s
went some way to compensate for this. They brought together the
multifarious sources of support for chemical knowledge. In the
1840s the association of academic and practical chemistry was
expressed through the foundation of the Chemical Society in 1841
and of a host of teaching institutions, pre-eminently the Royal
College of Chemistry, established in 1845. Examining the origins
and early years of the Chemical Society and of the Royal College of
Chemistry we see the emergence of a hierarchy of practitioners and
of the contents of the discipline of chemistry as it was shaped
through the activities of the early-Victorian chemical elite.

The Chemical Society's founders

The Chemical Society was not founded to further some great
national project. Instead, it was cast as a service to its members. This
service approach to the society was reflected in the retirement speech
of the first Secretary, the London professional chemist Robert
Warington, in 1851.[4] He gave three reasons for his participation in
the formation. His objectives had been firstly 'to break down party
spirit and petty jealousies', secondly 'to bring science and practice
into closer communication', and thirdly 'to bring the experience of
many to bear on the same subject'. Each of these objectives was
relevant to the interests of one or other category of founders.
Professional chemists squabbled over the merits of particular
products and processes, while academics scorned each other's
theoretical positions. The integration of science and practice was
desired by the academics for ideological reasons. Others had a

practical need for, and a worthy interest in, greater knowledge of science, not least because of its conferral of professional status. Finally, public discussion would assist in the resolution of the diverse problems faced by members. These varied objectives were aired at the founding meeting on 23 February 1841. Warington suggested that the the Society should offer:

> The reading of notes and papers on chemical science (in the most extensive meaning of the term), and the discussion of the same. The formation of a laboratory, in which might be carried out the more abstruse and disputed points connected with the science. The establishment of a collection of standard chemical preparations, of as varied a nature as possible, for reference and comparison, and thus to supply a very great desideratum in a metropolis; the formation also of a library, to include particularly the works and publications of Continental authors.[5]

These proposals, presented as a coherent whole, in fact spoke to a range of interests. The use of the term 'notes and papers on chemical science' could refer to any kind of communication and did not even specify originality. The proposed laboratory was presumably intended to resolve discord among the afficionados of esoteric knowledge and to provide what were seen to be increasingly expensive central facilities for individuals to use for their own work. The library of foreign publications was also for the use of the elite with interest in, and the ability to read, untranslated Continental authors. The proposal for a museum seems to have descended from an already defunct scheme, suggested in 1835, for Apothecaries Hall to establish a 'Museum of Materia Medica and Chemical Preparations'. By offering standards, such a museum would solve problems and disputes concerning the identity of problematic products.[6]

On the face of it, the four initial proposals were accepted and merely formalised by a sub-committee formed at the first meeting. In fact the committee subtly changed the emphasis: the Society would be primarily devoted to the discussion and publication of research.[7] The shift away from the vision articulated by Warington and reflected in the initial document seems to have been due to Thomas Graham, one of the young, activist academics seeking a reorientation of British chemistry. Elected the first president at the founding meeting, he envisioned not a variety of different facilities serving various interests, but a scholarly forum. Members of the audience would discuss, learn, apply chemistry and improve themselves, perhaps even to the rank of contributor. Hence the Chemical Society was not to be seen as having a completely different function from the Royal Society as first proposed, it would be just more specialised and more accessible, in fact a subordinate rung on the ladder of scientific achievement. In his inaugural address Graham first made appropriate symbolic reference to 'high chemistry', by mentioning the support of the two leading European chemists, Liebig and Dumas. He explained the plans for membership. There would be two grades. Members 'properly so called' would ideally be 'high' chemists — those men 'who have prosecuted the science with zeal and research'. Then there was to be a class of 'Associates' who would be without voting rights and who would not pay a subscription. The class would consist of 'a large number of young gentlemen pursuing chemistry as a science, such as pupils, managers of manufactories &c.'. At the same time as Graham had admitted all sorts of men into the constituency of the Society he had proposed a clear social hierarchy.[8]

Altogether thirty-one men were involved in the conclaves that led to the foundation of the Chemical Society in February and March 1841.[9] They included academic, professional, industrial and gentlemen chemists. Warington and Graham represented two of the important constituencies involved. Robert Warington was a professional chemist who had been apprenticed, had taught as an assistant and was a well known consultant in London. Along with him came the major established professionals, including his old master, J. T. Cooper, and W. T. Brande of the Royal Institution. Thomas Graham of course represented the new academic elite of the Thomson school. His fellow graduates, R. D. Thomson and Thomas Clark, though not London residents, moved a central motion at the founding meeting. The coalition between the two groups dominated the offices of the Society during its early years. By contrast the traditional philosophical elite was not represented. None of the surviving members of the Chemical Club joined. Faraday, by then doyen of English science, was not a founding member, refused the first presidency, and never contributed a paper. Daniell was elected to the first committee, but never attended nor did he ever present a paper. Manufacturers were present at the founding meeting but took a subordinate place in the Society's affairs.

The Chemical Society reflected the growing significance of the professional chemists in London and the aspirations of a new research-oriented body of young academics. This harmony was facilitated by their science's popularity during the 1840s. Because it seemed to offer knowledge that was both theoretically defined and potentially useful in practice, chemistry came to be seen as the very symbol of professional competence. It was promoted as part of the training for conventional professions and trades, and as the appropriate expertise for the solution of a wide range of problems. These expectations were exploited by many promoters of new institutions. Pre-eminent among these were the founders of the Royal College of Chemistry, which opened in London in 1845.

The Royal College of Chemistry: the proposers' vision of chemistry

In the first instance, the scheme to establish what became the Royal College of Chemistry was the personal project of two entrepreneurs with some expertise in practical chemistry. John Lloyd Bullock, a pharmacist, and John Gardner, an apothecary, saw the exploitation of the chemical issues of the period as a means of realising their own ambitions.[10] Education, Bullock preached, would bring not only status, but also opportunity, to the practical chemist.

There is a further use in thus becoming a good practical chemist; it will enable you to be the counsellor of the agriculturalist, of the manufacturer, of the physician — to spread the love and practice of chemistry.[11]

It was their belief that men with vocational interests in practical chemistry constituted a class who could share a common focus. Bullock and Gardner argued that chemistry was an applicable science, and they envisaged research on applied topics. They also foresaw considerable opportunities for profit (for themselves) through the exploitation of patentable discoveries.[12] Gardner and Bullock united a vision of chemistry as a science which was essentially useful with the interests of practical men. To realise this they proposed the division of scientific and of practical chemistry into two distinct but linked schools with different types of activity that should nonetheless be mutually reinforcing. One school would concentrate on principles and further research. The other would use such research and would focus on applications; the purely scientific would not figure there.

The idea of scientific chemistry employed by the two entrepreneurs was not original to themselves. They had adopted it from a pamphlet written by William Gregory, a Liebig pupil who was Professor of Chemistry at Aberdeen. In 1842 Gregory had published a polemical pamphlet urging government support for laboratory teaching on the German model.[13] This pamphlet had articulated views which were to be central not only to the Royal College of Chemistry but even to the whole subsequent history of British chemistry. Following Liebig, Gregory argued in a way that would become familiar:

... whether the object of the student be to qualify himself as a teacher of chemistry, to learn the bearing of that science on medicine and physiology, or to become a manufacturer, the same purely scientific education in the art of research is recommended to all. It would be impossible, for example, to teach specially all the different chemical manufactures. ... It is found by experience, that when all learn the general principles of chemistry, they acquire the special details of any manufacture in the manufactory in a far shorter time than they could have done in the laboratory ... even for directly practical purposes, the most purely scientific education is really the best, and is more certain to lead to improvements in practice than the most laborious experience in any one manufacture, gained as it generally is, at the expense of general principles.[14]

Not only was science primary, but research in it was vital to the prosecution of practice. Thus the position of the academic elite was assured. The argument went on to demonstrate that if science was the end, research was the means of education.

[technical advances] ought also to teach us, that diligent research after new truths, however they may appear at first remote from any practical application, will yield, in the course of time, practical results of equal or greater importance; that no well ascertained fact in natural science is ever barren: and that the best method of promoting practical improvements is to encourage scientific research.[15]

The original plan of Bullock and Gardner for two institutions collapsed; the Royal Institution, which was invited to house the pure science school, in the end did not accept.[16] However, support from several important men was raised, and a sub-committee launched an appeal for an independent college. They proposed to combine all the functions of chemistry advocated by Gregory. They wanted to establish laboratories for research and for teaching, and departments specially for the study of chemistry's many applications. Together with general publicity, special appeals were directed individually to sectional interests.[17] A professor from Germany, August Hofmann, a pupil of Justus Liebig, was hired to teach.[18] A building was occupied in Hanover Square and soon extended back on to Oxford Street near Oxford Circus. In October 1845 the College of Chemistry opened and was soon to win royal patronage.

Over 700 supporters had contributed to the finances of the new college. Medical men, chemists and druggists, agriculturists and manufacturers all supported the appeal.[19] The scale and spread of this support highlight a faith in chemistry as the solution to diverse practical and professional problems that would sustain the college in particular and academic chemistry in general. The reasons for this substantial interest in chemistry can best be explored through reviews of the principal interests represented on the list of subscribers.

Medicine

The best developed of the markets for chemistry to which the promoters appealed were the medical professions. For them the proposal for the college was especially timely, since the education of a new kind of doctor, the general practitioner, was then being formulated. The emergence of the general practitioner resulted in educational developments that followed Scottish precedents. Sentiment in favour of compulsory practical chemistry for medical students had developed in Scotland during the 1820s among those who felt that practical training was essential for understanding chemical science. In 1829 the Royal College of Surgeons of Edinburgh began to require of its candidates three months' training in practical chemistry. Keeping abreast of Scottish developments, in the same year, University College offered practical chemistry to medical students.[20]

A special London medical degree was established in 1837 when a reformed University of London was chartered. In favour of a medical qualification for GPs, reforming medical men had actively promoted the new organisation.[21] Though University College became part of the university and lost its autonomy, the new medical course was again based on the Scottish model. Basic instruction in the sciences underpinned modern practical training that helped define the special expertise of the University of London medical graduate. The examinations were to include laboratory operations. The one term of practical chemistry required by the syllabus was medically oriented: it included 'Practical Exercises in conducting the more important processes of General and Pharmaceutical Chemistry; in applying tests for discovering the adulteration of articles of the materia medica, and the presence and nature of poisons; and in the examination of Mineral Waters, Animal Secretions, Urinary Deposits, Calculi, etc.'[22]

Medical reformers argued that organic chemistry too should become part of the new curriculum. Although there was controversy about the fundamental theories of organic chemistry, the analytical methods developed for its investigation (primarily in the Giessen laboratory of Justus Liebig) were widely respected. In turn, the new methods began to be used for more systematic study of the processes through which organic constitutents were produced in both healthy and diseased living systems; organic chemistry came to be viewed as the key to the study of physiology and of pathology of both plants and animals. The reforming editor of *The Lancet* argued that a new London MD should be available for research, and in particular that the chemical investigation of physiological problems begged for the attention of researchers. However, there was a major obstacle to this programme: the lack of suitable facilities for studying practical organic chemistry.[23] Despite the importance of teaching for the London medical degrees, neither University College nor King's College had facilities for providing analytical and research training.

From 1841 another sector of the medical community, the chemists and druggists, also began to promote actively the study of practical chemistry so as to enhance their professional status. Traditionally they had been the the medical men for the poor; now the rise of the GP eroded their medical role. Worse, under pressure of competition, many chemists had become little more than grocers, selling a wide variety of goods.[24]

On the other hand, some pharmacists used their expertise at chemical preparation to diversify into small-scale manufacture. The denomination 'operative chemist' was taken by those whose work was chemical but more sophisticated than dispensing and routine preparation. They also sought incomes through chemical consultancy. These men were concerned to uphold their professional image.[25] Chronic difficulties became acute in 1841, with a

Parliamentary Bill that would have subordinated chemists and druggists to the control of their competitors, the apothecaries.[26]

The Pharmaceutical Society, established in 1841, had the immediate objective of fighting the legislative threat. Leaders of the Society admitted the need for professional control, but wanted to provide their own by means of pharmaceutical education. Their object was no less than the creation of a professional role for the English practitioner after the French and German model. Science would play a crucial role in this political process. By studying the sciences on which his art was based, particularly practical chemistry, the chemist and druggist would elevate himself from the role of tradesman to that of practising scientist, the pharmaceutical chemist.[27]

In pursuit of professional status, the founders endowed their Society with a number of roles. It was to establish a school, serve as an examining board, become the legal licensing body for chemists and druggists, and operate as a learned society complete with a journal for pharmaceutical subjects. The school opened in the autumn of 1842, though the well equipped laboratory for the practical course was only completed two years later.[28] Meanwhile, there were many arguments about the relative importance of chemistry and the other subjects proposed for the comprehensive curriculum. Lobbyists such as Bullock argued that the Society was offering too little chemistry, too late. Despite the foundation of the Pharmaceutical Society with its own educational goals, there was an unsatisfied need for chemical education for the pharmaceutical chemist.[29]

Hence it was argued by the proposers of the College of Chemistry that chemistry could enhance education in the new professions of general practitioner and pharmaceutical chemist. Through the study of chemistry, scientific, practical and professional goals were to be attained.

The landed interest

Equally the college's promoters exploited the landed interest's renewed faith in chemistry. Politically and economically landowners were still the most important force in the land. Both as agriculturists and as mine owners, they were coming to perceive a need for science. In 1838 they had founded the Royal Agricultural Society with the motto 'Practice with Science'.[30] Agriculture was becoming more businesslike. Increasingly, it has been said, farmers purchased raw material at the cheapest price and sold at the highest 'just like any cotton lord'. With changing economic roles were associated changing technology and the diffusion of 'high farming'.[31]

High farming entailed a variety of innovations. Important among these were costly fertilisers. From £600,000 a year in the decade after 1815, expenditure on animal feed and fertilisers increased to £3 million annually during the early 1850s.[32] Guano imported from South America was the new product that made the greatest impact. In 1841, 2,851 tons were brought into Great Britain. Four years later, in a peak year, 283,300 tons were imported. In addition, there were 'patent manures' such as lime from gas works, and superphosphates.[33] The introduction of expensive fertilisers was not without problems. It was difficult to tell how much fertiliser to use on a given soil and indeed whether the fertiliser was good quality. For answers to these questions, farmers turned to professional chemists, often at considerable expense. Moreover, as landowners were often also mine owners, aristocrats had interests in the ores beneath the ground.[34] Again they called on chemists for assistance.

From the mid-1830s, problems of land management spurred a number of institutional initiatives. As a result, in a new departure, several salaried posts for analytical chemists were created. The creation of the Geological Survey in 1835 had been an early, governmental response to the calls for the application of science to national needs. Though at first there was no chemical work by the Survey, in 1839 its Director successfully persuaded the government to spend £1,500 on an extension to the Museum of Economic Geology, which thereby created a position for a curator.[35] Richard Phillips, one of the senior professional chemists in London, was appointed at a salary of £200 per annum. In addition to his salary, Phillips charged up to two guineas for a private analysis. In his first eighteen months he managed to increase his income to £450 p.a. through analytical work.[36] Though his title was 'curator', Phillips had a variety of chemical duties. In practice, he was the government's own chemist. He was to serve as chemist to the Survey, analysing and classifying samples sent in by field workers, solving chemical problems for other government departments, performing analyses for private individuals, and lecturing on analytical chemistry, agricultural chemistry, metallurgy and mineralogy.[37]

Among the mainstays of Phillips' private work was soil analysis for tenants and landowners. The application of analytical chemistry to agriculture was heightened in 1840 by the publication of a much previewed translation of an important German work: *Organic Chemistry in its Applications to Agriculture and Physiology*.[38] The purpose of Justus Liebig, the author, was to explain agriculture theoretically in terms of organic chemistry, justifying at the same time his chemical theories. The possibility that this theoretical approach might eventually lead to changes in agricultural practice was of course important. The chemist, by identifying the composition of plants on the one hand, and the natural nutrients on the other, could dictate what and how much artificial fertiliser would be required. The book itself, however, was a programme for research, rather than a guide to good practice. Liebig stressed that much more research was needed before a routine procedure would be possible. For example, he suggested a nationwide series of plant-ash analyses that would look chemically at the same plants grown on different soils.[39] This plant ash scheme was taken up with Liebig's personal encouragement at the British Association meeting in September 1844. The British Association voted £50 for the project with the proviso (ultimately fulfilled) that the Royal Agricultural Society also provide funds. This discussion even succeeded for once in attracting country gentlemen to the Chemical Section.[40]

In Scotland, as in England, chemistry was seen as a key to the farmer's improvement. The old-established Highland and Agricultural Society, founded in 1784, in Edinburgh was soon lamenting the lack in Scotland of the cut-price analyses offered by the Geological Survey to the southern farmer. The landowners who ran the Society could find no money for a comparable scheme.[41] However, in 1843 a group of tenant farmers formed the Agricultural Chemistry Association in Edinburgh, hoping that chemistry would 'mitigate' the effects of low farm prices.[42] A year later, ten applications for the Association's new post of chemist having been received, J. F. W. Johnston, then Reader in Chemistry at the somewhat moribund University of Durham, was appointed Chemist to the Association with an income of about £400 a year.[43] The Association grew and prospered rapidly. By 1845 Johnston had a laboratory in Edinburgh and five full-time assistants who were paid about £300 a year; the Association had 764 subscribers and a thousand analyses had been completed. This example was followed closely elsewhere. In Ireland, the Chemico-agricultural Society of Ulster was founded in 1845 with the aim of assisting tenants.[44]

Difficulties in the colonies also inspired the appointment of chemists. For example, labour costs shot up after the emancipation of the slaves in British Guiana in 1838. A Royal Agricultural and Commercial Society was founded in 1844, and in 1845 the colonial government appointed its own Chemist, at a salary of £1,000 a year.[45] Where government did not help, private interests stepped in. Jamaica, like British Guiana, was hit by a crisis in the sugar industry. Despite their lack of resources, the planters founded a Jamaica Agricultural Society on the model of the Highland and Agricultural Society and of the Agricultural Chemistry Association. Two chemists were hired and sent to train under J. F. W. Johnston in Edinburgh.[46]

In addition to analytical service posts, there were also initiatives in the area of agricultural education. The possibility of the repeal of the Corn Laws increased interest in scientific means for improving production. Having weathered the severe depression of 1839-42, the landed interest feared the threat of foreign competition. Whatever the outcome of the Corn Law issue, it made sense to progressive farmers to advocate science. If repeal succeeded, scientifically increased production would improve England's competitive position; if repeal failed, scientifically increased production would satisfy domestic demands. Should the claims of scientific propaganda be realised even partially, investment in science would be amply repaid. Such a preparedness to insure through science was exploited by those seeking support for agricultural education.[47] In 1841 Charles Daubeny, who had drawn upon his Scottish medical education to acquire three Chairs at Oxford (chemistry, botany and rural economy), urged the Royal Agricultural Society to take more interest in agricultural education. The professor pointed typically to Continental, especially French, educational institutions. He suggested that, lacking similar support, English agriculture was not keeping pace with foreign advances.[48]

Citing Daubeny's call, in 1843 a group of private individuals set about establishing an agricultural centre. The college which opened two years later at Cirencester was to give a quasi-professional education to a new generation of 'agriculturists'. In addition to a practical department, the college was to have an academic side. This would feature courses in aspects of science ancillary to agriculture, including practical soil analysis.[49] Daubeny hoped that the centre at Cirencester would have research as well as educational functions. He envisaged a network of similar agricultural schools strategically sited on different geological formations undertaking national analytical projects.[50]

The extensive section of the Royal College of Chemistry proposal directed at landowners was carefully worded. Although no promises were actually made, the proposal's rhetoric may have encouraged the agricultural reader to believe that an analyst would be immediately useful to him in fighting foreign competition. Moreover, underground resources would be uncovered, revealing wealth in minerals and domestic phosphate fertilisers. These suggestions were hardened in a supplementary pamphlet that did make promises of a national geochemical survey of both soil and subterranean resources. This survey would be carried out by a network of provincial pharmaceutical chemists, who would be called upon to perform the analyses. In this way the college could be the point of contact between agriculturists and pharmacists.[51]

Manufacturers and engineers

Practical chemistry was thought to bear on several contemporary interests of manufacturers and engineers no less than on interests of landowners and medical men. Indeed, some scientific propanganda of this period held up the use of chemistry by manufacturers as a model for both agriculturists and chemists and druggists.[52]

There were already many areas of manufacture in which chemistry was used effectively at this time. Some highly innovative techniques, such as those used in metal refining and in electroplating, did make use of chemical expertise. Certain industries such as brewing and soap manufacture operated on a scale that necessitated process control. In large breweries the progress of temperature, specific gravities and times were all carefully monitored and recorded and chemical advice was available from independent professional chemists. Since only a few firms employed their own 'chemist', even in large enterprises proprietors, managers or other non-specialist employees were expected to possess some chemical expertise. In 1838 James Young, an assistant of Thomas Graham with no industrial experience, was hired to manage a factory of the great Liverpool industrialist James Muspratt. Young reported to his mother:

He [Muspratt] said if I had gone to his work ... six months ago I would have saved him five thousand pounds that had been spent on a bad patent while he was away.[53]

As a proprietor in the soda industry, Muspratt had what was still unusual faith in scientific training. However, in the calico printing industry all the top firms employed chemists. This industry operated at the fringe of the largest of large-scale businesses, the cotton industry. Chemists were needed because waste was expensive on so large a scale. Moreover chemists had an important role in innovation for commercial advantage as superior pattern, form and style were made possible by the use of new dyes and new methods of bleaching.[54]

In January 1840 a new journal made its appearance, in an industrial context of ebullient growth, competition and social flux. *The Chemist* was edited by the otherwise obscure Charles Watt, 'Lecturer in Chemistry', and John Watt. They intended that the journal should serve the distinct communities of chemical manufacturers, pharmacists, and chemical analysts.[55]

In the opinion of *The Chemist*, knowledge of chemistry was fundamental for the professional, but the journal designated research as totally alien to his concerns. The first volume referred contemptuously to those few magazines 'composed of dull and heavy subjects of abstruse calculation, written expressly for those journals, and which seem as if they were intended to gratify the inclination of the authors for displaying their profundity, and to limit the number of their readers'.[56] This formulation reduced the three chemical communities to one centred on the man concerned to apply science. While the three interests served by *The Chemist* did have their own individual concerns, and the journal was consequently divided into parts dealing respectively with chemistry, manufacturing and pharmacy, certain fundamental objectives were seen to be common. One, which was given great importance at first but then less stridently pushed, was the reform of patent law. The improvement of chemical education was a more persistent issue. By 1844 *The Chemist*, was reporting — one wonders how accurately — that:

The worthy manufacturers of the generation now passing away are having their sons instructed in chemistry, as affording the best means of competing with others, and of protecting themselves from the machinations of the fraudulent.[57]

The 1840s saw the rise of new engineering specialties — metallurgical, mining, and gas engineering — whose practitioners wanted to put their work on a more scientific basis. Despite its commercial importance British metallurgical practice was for the most part carried on without resort to science. During the early 1840s it was argued again that unless production was improved through the use of science, Britain would lose its resource-based industrial lead and become dependent on foreign products.[58] Many of the still intractable technical problems required chemical resolution. It was also argued that science should be applied to a humanitarian problem made prominent by mining disasters in 1842 and 1843. Government reports on these disasters emphasised the necessity of scientific education for professional men connected with mining, and they compared education provision in Britain unfavourably with that on the Continent.[59] An industry that appeared to the public even more potentially hazardous, coal gas manufacture and distribution, suddenly entered a new period of expansion during the early 1840s.[60] Three new companies appeared in London in 1842 and 1843. The industry, in search of expanding markets, concentrated on making gas suitable for domestic lighting. The problems involved were mainly chemical; a thorough knowledge of organic chemistry, which was just beginning to sort out the products of the distillation of oils and coals, became increasingly important to the gas engineer. An enquiry into faulty gas mains in the late 1840s recommended that every gas works employ a chemist, both for its own benefit and for that of the public.[61]

Engineering was developing into a profession and the number of English engineers was growing rapidly. For engineering as for medicine, the promotion of strict academic entry requirements was a means of affirming professional status. The formal study of applied chemistry and of other sciences was directly useful to prospective engineers and it also helped define their professional expertise.[62] From the late 1830s, a number of opportunities arose for manufacturers to take courses in applied science. Most of these courses were part of programmes, all of which included chemistry, for training engineers. From 1838 there were formal engineering courses at King's College London and at the University of Durham; the Putney College for Civil Engineers opened in 1839.[63]

In their appeal to manufacturers, the proposers of the College of Chemistry again stressed the importance of science in relation to the struggle against foreign competition.[64] As a specific example of the potential of chemical science, the proposers drew the attention of manufacturers to Liebig's discussion of the sulphuric acid industry in his *Familiar Letters*. There, Liebig had shown that sulphuric acid was a crucial industrial chemical and that the principal current method of manufacture had been discovered by scientific investigation. The proposal pointed out that within the industry there were still many chemical problems such as how to substitute indigenous sources of sulphur for foreign ones.

A special supplement to the proposal drew on the support of David Mushet, a founder member of the Chemical Society, well known for his important innovations in the manufacture of iron and steel. Including a lettter from Mushet, the supplement was addressed specifically to the proprietors of mines and metallurgists. The appeal was cast in imperial, not merely national, terms: 'an incalculable amount of mineral wealth exists in Great Britain and its Colonies, and also in India, concealed from its proprietors only for want of knowledge'.[65] It stressed that the application of chemistry to the problems of mining and metallurgy would forestall foreign competition. Chemists might develop extraction and refining processes for known, abundant, domestic sources of metallurgical raw materials that were previously imported. Furthermore, research might be done on possible uses for metals, such as tungsten, that were known to be in abundant supply, but which were not mined because no one knew what to do with them. Chemical investigation of metallurgical processes might also reveal, in materials traditionally considered waste, products valuable for other purposes such as agriculture. All these examples were selected to direct attention to the importance of training chemical investigators as well as routine analysts. Thus the need for the professional chemist was made evident from yet another quarter.

The bulk of the financial support was derived from the landed interest. However, each of the constituencies came to be important. Medical men, for example, undertook most of the administrative responsibilities.[66] Students were to come from each of the interests, though soon aspiring manufacturers came to predominate, replacing medical and pharmaceutical students [67] Research problems were submitted from all quarters.[68] Therefore, as the college needed to be sustained as well as built, professional interest and support were continually important.

The impact of the new pedagogy on the chemical community

The imminent establishment of the Royal College of Chemistry galvanised the existing London colleges, King's College and University College, into building laboratories and instituting analytical courses.[69] Since at neither institution were the professors willing to take on the additional labours, Giessen-trained assistants were promoted.[70] The commitment to training practical men and professionals was common to the new laboratories. At University College special classes were held in the evenings so that manufacturers could attend.[71] Even outside London comparable developments could be found. In Birmingham another student of Liebig particularly emphasised the value of his course to medical students.[72] In Liverpool, with thirty students, Sheridan Muspratt, a Giessen PhD, obtained accreditation from the Society of Apothecaries.[73] The support for chemistry from diverse constituencies brought about a new small chemical professoriate. A numerically small but historically important proportion of students in the new colleges, particularly the Royal College of Chemistry, reached out to academic careers.[74] There was neither a tradition of leading an academic career in chemistry, nor means of doing so. However, the formation of new colleges and the emergence of new posts began to make such a career possible.

New opportunities for chemistry teachers also emerged in Ireland, India and Canada. In 1848 three non-sectarian colleges were opened in Ireland – the Queen's Colleges at Belfast, Cork and Galway. Each employed a Professor of Chemistry.[75] In India a professorship of chemistry was established at Madras Medical School in 1846. John Mayer who studied at the Royal College of Chemistry was an early appointee.[76] In Canada the University of Toronto opened its doors in 1842 and appointed a Professor of Chemistry, applying to Faraday for suitable candidates.[77]

The value of research

Although today we may associate research with university teaching there was no necessity for the incumbents of the new positions to publish. Nevertheless, at this time research did come to have vocational importance. This happened in Germany as State bureaucracy sought rational ways of selecting university teachers.[78] Less formally than in Prussia but all the same in an analagous manner publicly recognised research competence, and hence publications, became a useful qualification in Britain for getting one of the scarce jobs.[79] The growing importance of research credentials is evident in the history of University College. The first Professor of Chemistry appointed there in 1828, Edward Turner, was known for his textbook rather than contributions to knowledge. However, when the Senate Committee appointed his successor, Thomas Graham, it was his European reputation that weighed most heavily. Still, it was important that he had had teaching experience at the Andersonian Institution and a delegate was sent north to check on Graham's lecturing ability.[80] Twelve years later, in 1849, research excellence was the sole criterion in the appointment of a Professor of Practical Chemistry. One candidate, who had been the assistant in the laboratory for several years, and in fact had conducted classes himself during the sickness of the previous professor, Fownes, was discounted because he had not 'yet published any original work on chemistry'. John Percy, who had taught chemistry at Queen's College Birmingham, was a strong candidate with five years' experience of teaching practical chemistry. But the Senate Committee found that Percy, with a medical background, had conducted too little peculiarly chemical research.[81] Finally, Alexander Williamson was appointed without teaching credentials but with a fine research reputation.

As an incentive to research, the perceived importance of a scientific reputation was perhaps more important than its actual utility in getting a chemical job. The correspondence of the ambitious and politically astute Lyon Playfair is particularly replete with reflections on the best means of advancing. In 1842 Playfair, fishing for a position at the Museum of Economic Geology (which he failed to get then), wrote to Liebig:

I *must* publish some original paper in Agricultural Chemistry, in order to give me some claim for the appt regarding which I once wrote to you about.[82]

In June 1843 Playfair's friends thought that his research record was insufficiently substantial. The calico printer John Mercer wrote encouraging him to publish 'but not in such small bits as Croft, Warington &c but a regular investigated essay'.[83]

The autobiography of Playfair's pupil, Edward Frankland, contains a passage describing thoughts over whether to marry as he completed his PhD in Germany. It captures the hope fostered by a research reputation in the 1840s.

In fact I had nothing to live upon except £70 per annum allowed me by my parents. Would it not then be heartless and cruel to try to persuade Sophie to link her fortune with a penniless student? On the other hand, I was hopeful and ambitious. I had already acquired somewhat of a reputation as a discoverer of ethyl, a discovery which at this time was exciting much interest in the chemical world, and was communicated to many learned societies and announced in many journals and periodicals. When in an optimistic mood, I could not believe that this belauded discovery, together with the training I had received in chemistry and allied sciences culminating in my university degree [PhD from Marburg], would not entitle me to obtain some professional employment with a 'living wage'. On the other hand, when in a pessimistic mood, I remembered numerous instances in which discoveries of far greater value and interest had, so far as their authors were concerned, been passed over without any substantial reward. My optimistic moods, however, were more frequent and prolonged than my pessimistic ones, and so it came about that after a sleepless night spent in anxious thoughts, I resolved to ask Sophie to be my wife on the next occasion of our meeting.[84]

Frankland had decided to marry because he had judged, correctly, that research achievements could be the basis for his career. His subsequent appointment to Owens College fulfilled the novel expectation of an academic future.

The impact of the Chemical Society on the new chemical community

A consequence of the new academic career was an unanticipated distancing of young academic chemists from practical contexts. Research became their prestigious vocational activity. The new educational routes from academe to practical careers subordinated the practical to the academic chemist. The status of the academic and of research was thereby elevated.

The primacy of research was fostered by the Chemical Society: most important of all its activities were the publications. More money was spent on these than on any other activity.[85] In 1847 a regular *Quarterly Journal* was established and a publications committee appointed.[86] Its members were the major London professors: Graham, Hofmann, W. A. Miller from King's College and Lyon

Playfair at the Geological Survey. In this was reflected the growing importance of the younger generation of chemists. Five of the eight members of the sub-committee that recommended the decision had academic ties with University College and the Royal College of Chemistry. Only two months earlier Hofmann had given up attempts to establish his own journal at the Royal College of Chemistry. In practice the *Quarterly Journal* which he and his colleagues proposed was a substitute for that project. It enabled his pupils to find a ready home for papers that demonstrated their research competence.[87]

Authors

At first the Society's publications were dominated by the founders.[88] Graham, Warington and the Scottish analytical chemist, John Stenhouse, each contributed a dozen or more papers over the period and together were responsible for a quarter of the papers recorded in the first volume of *Memoirs and Proceedings*. Contributors of their generation characteristically combined the theoretical and the practical by producing papers on both. Graham, for instance, read thirteen papers to the Chemical Society during the 1840s. Some reported on his long-run research programme on heat produced by chemical reactions, others reported diverse observations such as on phosphorus contained in well waters (possibly of interest to agriculturists), the hypochlorite in the refuse from gas works (of interest to the new photographers), on the composition of fire-damp, and on the mode of obtaining iodine from the kelp of Guernsey.[89]

By the second half of the 1840s the journal came to be dominated by the rising new generation of professors and their students and assistants. Giessen, University College London, and the Royal College of Chemistry were well represented. The character of the papers reflected not only the desire to communicate knowledge but also personal ambitions. The contributions of the majority were simple analyses which served the purpose of demonstrating the author's research competence – such were the series of mineral water analyses and investigations into homologous compounds published from the Royal College of Chemistry.[90] By contrast to routine work, there were also brilliant papers, some of which were strikingly self-conscious. In 1845 Muspratt and Hofmann presented a paper to the Chemical Society on the transformation of toluol to a new alkaloid, toluidine. The introduction by the two authors, both in their late twenties and assistants at Giessen, portrayed the work as both a necessary step in the internal development of chemistry and as potentially of enormous practical importance. It began with the reflection that while analysis had been the traditional mode of organic chemistry, synthesis would have much greater theoretical and practical importance. For example, 'if a chemist should succeed in transforming in an easy manner naphthaline into quinine we should justly revere him as one of the noblest benefactors of the human race'.[91] Their paper was a step in that direction. Three years later Liebig rewarded Muspratt's contribution with a glowing testimonial in support of his new Liverpool College of Chemistry: 'The results which he [Muspratt] has furnished must be regarded by all chemists as real treasures and I am convinced that Dr Muspratt would perfectly fill the office of Professor.'[92]

The link between intellectual and professional ambition exemplified by Muspratt was apparent also among other contributors to the proceedings of the Chemical Society. The papers of Frankland and Kolbe, published by the Chemical Society between 1847 and 1850, were self-consciously path-breaking. Indeed the authors apologised for their apparent hubris. As Frankland recorded, their work was highly acclaimed when read to the Chemical Society.[93] And as Frankland's autobiography shows, he was well aware of the relationship between an outstanding research reputation and career prospects as a chemist.

Given their importance in the foundation of the Society, it is striking that the older professional chemists did not put their imprint on the discipline through the journal. However, their careers had not entailed the publication of research. Nor, despite their rhetorical importance, did manufacturers comprise a large proportion of the authors.[94] Of course published research had no direct career interest to them. Manufacturers should, however, not be seen as only interested in the chemistry of the processes in which they were commercially engaged. James Joule for instance, though a brewer, made fundamental contributions to science. All the same, most papers contributed by men working in industry did relate to the problems faced in the practice of their work. John Mercer and Walter Crum contributed papers on the chemistry of dyeing and of bleaching respectively.[95] Sometimes the significance of an industrial contribution transcended the particular context from which it had sprung. The outstanding example was the research reported to the

Chemical Society by James Napier and Charles Glassford, employed
by Elkington's, on the cyanides which had been discovered to be the
appropriate electrolytes in plating processes. This work, presented in
a series of eight papers, was the most sustained effort at applied
chemistry to be found in the Proceedings.[96] Such contributors were
neither numerous, nor typical of the manufacturers generally. Their
links with the academics were unusually close, and their participation
in the Society was much above the average.

So the academics, despite their nationally small numbers,
dominated the published Proceedings of the Chemical Society even in
its early years. At the same time the Society aimed to bring 'science
and practice into closer communication'. The simplicity of this
expression obscured the fundamental differences in the way priorities
were assigned and problems classified in 'science' and 'practice'.
Contributions to science could be evaluated in terms of their relevance
to the problems of particular specialties, such as organic chemistry,
and their compliance with the norms and standards of those
specialties.[97] Contributions to practice, on the other hand, could be
classified according to the industrial problem with which they were
concerned and evaluated according to their relevance to these
problems.

The majority of papers, almost all by academics, concerned
problems internal to chemistry. This genre was exemplified by some
of the papers on organic chemistry in the first volume of *Memoirs and
Proceedings*. There were four papers, all by students of Liebig,
devoted to supporting Will and Varrentrapp's new method of
nitrogen analysis against the aspersions of the Frenchman Reiset.[98]

There were also papers given significance primarily by technology.
The characteristic form was an explanation of chemical processes in a
manufacturing industry, or of the properties of material with actual or
potential economic significance.[99] There were very few papers that
directly suggested improvements in industrial processes. Perhaps the
economic advantages of secrecy outweighed the prestige of
publication. The difference between the two genres was reflected in
papers dealing with gun-cotton, which was a topical subject in 1846.
Five papers on the newly discovered substance were read to the
Society within eight months. Professionals and manufacturers were
interested in its composition mainly because of its explosive power
and the implications for the nitric acid business.[100] By contrast, a
young student at University College analysed what he called

'pyroxylin' as an attempt to identify quantitatively the results of the interaction of nitric and sulphuric acids on cellulose. Only in the title of his paper did he identify the subject of his analysis as gun-cotton.[101] Since the academics with their career interests in research dominated the journal, the resulting image of chemistry was academic and specialty oriented.

Conclusion

The Society created in 1841 was not envisaged exclusively as either a means for the advance of science, or for the understanding of chemical practice; it was intended to be both. The charter granted in 1848 boasted the intimate relationship between science and industrial practice in chemistry. The purpose of the Society was:

the general advancement of Chemical Science, as intimately connected with the prosperity of the manufactures of the United Kingdom, many of which mainly depend on the application of chemical principles and discoveries for their beneficial development, and for a more extended and economical application of the industrial resources and sanatory condition of the community.[102]

The publication of research soon came to be the chief purpose of the Society. That research was increasingly dominated by academics whose work sought to advance specialty interests. It enabled the Society to gather under the discipline of chemistry an enormous diversity of subjects.[103] The question of how this diversity was to be translated into a pedagogical system had had to be faced in the 1840s by schools such as the Royal College of Chemistry. Like the Chemical Society at its foundation, the college was intended to have a dual function to advance both science and practice. The college aimed to satisfy the many practical and professional interests which suported it by promoting a common discipline of chemistry. Practical skills which could be used in many contexts would come from research in that discipline. Research training became the focus of chemical education and, as a result, research in itself acquired primary significance. Academic chemistry was still wanted as the basis of practice. Nonetheless, as its diversity of practical aims led to a focus on research, academic chemistry came to be abstracted from practice.

Chapter Three:

Visions in institutional form

The coalitions of the 1840s were expressed not only through the building of institutions. They were also manifest in the way that an academic discipline of chemistry was defined and promoted through those institutions. The academics had a vision of a self-contained discipline which they articulated through the curriculum. At the same time the realisation of this curriculum relied on convincing the industrial and professional interests, on which their instititutions depended, that chemistry would be useful. Hitherto framed in terms of undifferentiated 'research', the chemists' argument came to be based on an explicit division of labour between complementary categories of 'pure' and 'applied' science. The academic vehicle for the promotion of these categories was the curriculum that had been devised at the Royal College of Chemistry in the 1840s and whose pattern dominated chemical teaching till well into the twentieth century.

The Royal College of Chemistry

Early in 1846, as the college began the second semester [sic] of its existence, the Secretary, the entrepreneurial John Gardner, spoke to an audience of actual and potential subscribers.[1] He descibed the official vision of the education that was being offered. The student would begin by obtaining a preliminary 'speculative' acquaintance with the existing body of chemical knowledge, either through reading available texts beforehand or by attending introductory lectures. He would then proceed to the laboratory to begin '... a course of experiments with his own hands, which produce the actions or manifestations which are characteristic of every known substance, at least, all of which are of ordinary occurrence'. It was emphasised that even though the student would be following a laboratory manual at this stage, he would profit from working at the college rather than independently at home. The college could offer him the constant supervision of a well trained teacher as well as the inspiration and assistance of fellow students at all levels.

Once the student had mastered all the known preparations, properties and reactions of the known substances, he would put his new knowledge into practice by undertaking a series of qualitative analyses of unknowns. The professor was to assign the unknowns in order, from the simplest to the most complex. The student would start this phase of his training by analysing a series of twenty solutions, each containing one base; he progressed to another series of twenty solutions, each containing one acid and one base. Then he analysed a series of solutions containing two or more bases plus two or more acids; ' ... in the end, all the inorganic bases and a number of acids in admixture are analysed to try the student's memory'. Only after mastering qualitative analysis was it assumed that the student would be ready to move on to quantitative inorganic analysis; the culmination of his training was seen to be organic preparation and combustion analysis. The Secretary perhaps rather optimistically described this last step as 'comparatively easy'.

Not long after the talk describing this curriculum, the first professor at the college, A. W. Hofmann, formalised the initial phases of study by offering two introductory lecture courses on inorganic and organic chemistry. Hofmann estimated that an average student should be able to complete the qualitative and quantitative analytical programmes in one year of (presumably full-time) study.[2] Although he did not explain how long a student should expect to participate in the college's research programme before being considered a qualified research worker, there is some indirect evidence. Nominally the college awarded a 'Testimonial of Proficiency' to signify both successful completion of the analytical course and the production of a piece of research of publishable quality; the one student who actually received this diploma required three years of full-time study. Since he was the talented William Crookes, three years was probably not considered excessive. The ideal of the Royal College of Chemistry education followed closely the curriculum at Justus Liebig's famous laboratory at Giessen, where Hofmann himself had studied.

The practical audience addressed by the college's Secretary in 1846 had to be convinced of the educational value of the curriculum; the talk was a propaganda exercise as well as an exposition of the content of the course. Gardner put across the image of research training in chemistry that had been pioneered in Germany as the way of giving both of the kinds of education traditionally desirable in Britain. On the one hand it could provide the liberal education which trained the mind, generally the preserve of the classically educated upper classes.[3] On the other, it offered directly useful vocational training otherwise accomplished by apprenticeship. Research training, Gardner declared, would be given by means of the analytical problems which obliged the student to 'become acquainted with the principles and theories of chemistry in order to draw correct inferences from the phenomena he creates ...'.[4] The student who underwent this sort of training would be very well prepared to undertake research subsequently in any field, scientific or practical. This ability was a state of mind as well as a technical competence. In other words, Gardner argued, through instruction in chemical theories, facts and techniques of manipulation, the student acquired a 'rigid mental discipline'. This outcome, he emphasised, was precisely the aim of an education at Oxford or Cambridge.

The deep study of classics and mathematics [the core of Oxbridge education] teaches men *how to learn*, as well as furnishes them with the instruments available on all occasions and on all subjects, and the facility with which the mixed sciences are acquired by those who have passed successfully through our universities is a matter of common remark.[5]

Gardner further extended the comparison between the college and the old universities by suggesting that the professor's close personal supervision of advanced students in the laboratory was equivalent to the tutorial system. The lecture triumphantly concluded with the three benefits to be gained from the college's chemical curriculum. To have learnt analysis was to have acquired a powerful tool: 'for advancing one's own knowledge, for practising the useful arts, and manufactures, and for extending the boundaries of science'. Such linking of the German research tradition with English liberal education and vocational training was an effective argument. The gambit came to characterise the strategies of those wishing to promote the study of science.

Despite the virtues of the ideal course, in practice very few students followed it. The curriculum was modular and highly flexible. Students could enroll for the lectures only, as well as for any number of hours in the laboratory up to a set maximum for each of two twenty-week terms (semesters) each year. Thus not only could each student proceed at his own intellectual pace, as the college Secretary had promised, but he could also fit chemical study into his other commitments. In the early years almost 80% of the first 356 students attended for only two semesters or less; roughly 50% left after one semester. Attendance of that '80%' was not full-time – on average such a student came for only 60% of the time available during his one or two semesters. This attendance pattern persisted into the 1850s and '60s. On Hofmann's own estimates, this sort of attendance would have enabled average students to gain only rudimentary acquaintance with analytical techniques. It could hardly warrant the claims to liberal education and professional training, not to mention the advancement of the frontiers of knowledge.

There is some fragmentary evidence about students' expectations at the time. The father of one young prospective student wrote explaining his requirements.

My son is very anxious to devote a little time to the study of chemistry at one of the London institutions with a view to its application to brewing; he cannot possibly be spared from the brewery for more than 8 or 10 weeks; I shall be greatly obliged if you will inform me what will be the terms for such a period at the London University and at what time between this and September he would derive most benefit.[6]

The college's flexibility enabled the young man to attend full-time for one semester during each of three years. It is unlikely that while there he would have studied anything so specific as the chemistry of brewing; however, often students' own analytical projects were in a practical area. In 1848, for example, one student, who was a brewer's chemist, systematically analysed a large range of British beers.[7] Another student spent three days per week at the college during one semester to prepare for the Indian service. He admitted that he had neither aptitude nor interest in chemistry. He merely wanted to acquire a few analytical techniques needed for his posting.[8]

The flexibility was also featured as an attraction to aspiring professionals. Some 19% of the original group of students destined for medical or pharmaceutical careers attended for only one or two semesters. For medical students, who all had to study chemistry, it was normal practice to shop around for the best and most convenient 'buy' amongst courses which were approved by the qualifying body of their choice. After 1853, when the college was linked with the new School of Mines, hardly any medical students attended the chemistry course. The emulation of the college by institutions such as University

College with its more comprehensive medical links had obviously been successful. Equally the School of Pharmacy seems to have competed successfully. Nevertheless, flexibility attracted enough students to the college to ensure its viability.

The flexible curriculum was also important to that minority wishing to pursue research. As at Giessen, a small number of advanced researchers came on an 'occasional' basis to work in the laboratory. Though they constituted but a small proportion of the student body, the research workers gave the college a lasting reputation which came to overshadow the memory of other functions.[9] This role of the college as a centre of advanced research was actively fostered by the professor, August Hofmann. His priorities were indicated when supporters tried to take up an early promise that the college would give them cheap analyses. Hofmann was reluctant to oblige. Similarly the college Council expressed disappointment in 1851 when Hofmann turned down the post of non-resident assayer to the Royal Mint, estimated to be worth about £300 a year. But, Hofmann wrote to his mentor Liebig, '... I do not want to become an "analysis machine", because I have research plans for the next ten years'.[10] On another occasion, having just spent six weeks doing a chemical study of London's water supply, Hofmann again lamented his position to Liebig:

Since here [in England] knowledge is thought of and esteemed only in its relation to life, so one cannot easily avoid tasks which pertain to practical questions of life, especially since these will be highly honoured. One earns money this way but is drawn unawares into another field in which he has not the same interest or success.[11]

Thus despite the college's public rhetoric, to his academic mentor Hofmann professed disdain for the merely practical applications of chemical skill. It is ironic that the aspect of application that he found least worthwhile was the task of routine analysis, at the very focus of the training which he gave to his pupils. The research which in the supporters' eyes was only a pedagogical vehicle, or possibly a valuable route to useful discoveries, was to the professor the principal objective of the practice of chemistry.

University College London

University College had been the established leading centre for chemical training in the metropolis. It was alarmed by the threat of competition from the glamorous new college, and immediately established a new Chair of Practical Chemistry. Thomas Graham, the Professor of Chemistry, was, of course, in principle in favour of practical teaching. He was one of the long-standing spokesmen for the essential utility of chemistry. He had been an architect of the Chemical Society and an author of a popular textbook. At the same time he was an active researcher who laid the foundations for important steps in both inorganic and organic chemistry, the theories of osmosis and of colloids. Nevertheless, Graham was not attracted to the time-consuming labour of teaching analytical chemistry.[12] His main task at University College had not been to train chemists, though he did take one or two private pupils a year. For Graham, in common with other London-based academics, the bread-and-butter teaching was in lecturing to the medical students.[13] He construed the following as satisfying medical requirements in 1838:

The student will be exercised in conducting processes selected from all the departments of chemistry, & in the manipulations of testing and analysis. He will have an opportunity of becoming acquainted with many mineral substances used in the arts particularly the metals and their ores, & receive instruction in assaying.[14]

So, even before the establishment of the College of Chemistry, Graham stressed general training in inorganic chemistry, which he had previously given to aspiring manufacturers as well as medical students in Glasgow.[15] On the other hand, he complained about the expense of teaching practical chemistry and the incursions on his research time which long spells in the teaching laboratory would entail.[16] When the University of London established its own medical curriculum in 1839 with its stronger component of vocationally oriented practical chemistry, Graham's views were certainly reflected in his college's response.

The clause describing the course of practical chemistry ... seems to us more particular than is necessary, and might possibly induce some teachers to proceed too hastily to those operations which come into use in medical practice, to the neglect of the more elementary exercises, a familiar knowledge of which is the only sure foundation for a knowledge of the other.[17]

In the event, Graham managed to satisfy his own interests and the requirements of the university by devoting a portion of his laboratory course to organic chemistry.

There were financial as well as intellectual reasons for Graham's caution. When the establishment of the College of Chemistry was being discussed, the possibility of attaching such a teaching programme to one of the existing London colleges was mooted. For himself, Graham made it clear, he 'had no wish to be encumbered with the new college'.[18] Laboratory teaching of chemical analysis was both unremunerative and time-consuming. Graham suggested that the Scottish system used at University College of payment of teaching staff by capitation was partly at fault; it favoured the large lecture class over the small teacher-intensive laboratory class. This was especially true of practical chemistry, where there was the additional expense of providing chemicals and apparatus out of the professor's share of the fees. It was doubtful whether a fee could be found that was both low enough for the students but high enough for the professor. In general Graham warned that if a professor did not earn enough from his classes, he would have to take on extra duties to make a living; that would erode both his own research time and the time available to his students.[19] Just as Hofmann later discovered, unless they were supplemented, the professor's resources would be unduly stretched by the competing demands of high research and routine analytical teaching.

It is hardly surprising then that when University College opened its Birkbeck laboratory in 1846, Graham took no part in it. Teaching there was carried out by his erstwhile assistant, George Fownes, now appointed to the specially created Chair of practical chemistry, and the college subsidised the programme. Here too the course concentrated on the self-contained discipline rather than on its applications. And just as at the Royal College of Chemistry, a good deal of time had to be spent on the basics.

The *University College Calendar* for 1853-54 indicates how the course had developed. Chemistry was part of the curriculum for the BA (Hons), for the course in civil engineering, and for medical degrees. Despite the diversity of their interests, the same lectures and laboratory training were given to the three groups of students. In terms similar to those used in the rhetoric of the College of Chemistry, University College advertised its laboratory course as useful for mental discipline, professional education and practical application. Again, the laboratory was open for similar hours and the fee levels and flexibility were the same as at the Royal College of Chemistry.

The lecture course given by Graham consisted of a daily hour-long lecture for two terms, inorganic chemistry for the first term and organic chemistry for the second. The subjects covered were predictable: heat, light, properties of the non-metals, atomic theory with affinity and electricity, the metals, vegetable substances and their changes, animal substances and their changes. As in Graham's influential textbook, there was rhetorical obeisance to the teaching of chemistry in its scientific and practical contexts. However, in practice, applied topics were taught to the level of gentlemanly familiarity only, and then, with the exception of certain medically relevant techniques, only at the level of principles. The *Calendar* proclaimed:

In discussing chemical laws and the properties of bodies, their bearing upon the economy of nature, and their useful applications in the arts will be insisted upon. Hence it will be a prominent object of the Course to develop the principles of important chemical manufactures, such as glass-making, the working of metals, gas-making, bleaching and dyeing, calico-printing, brewing, distilling, and the preparation of various chemical products used in pharmacy. The manipulations and practices of testing will also be exhibited and applied, particularly in the detection of poisons, and of adulteration in the case of various chemical products.[20]

Again, University College civil engineering students took the same inorganic lectures as the general and medical students, and therefore learned about the most important chemical manufacturing processes in no more detail. Although the civil engineers attended the same organic course as the others, the section of the *Calendar* directed particularly to them stressed the practical orientation. For example under the heading 'composition of organic substances' Graham included references to a variety of fuels, manufacture of vinegar, fermentation, brewing, distilling, and the action of antiseptics. The heading 'ordinary constituents of plants' included the baking of bread, varnishes, paints and the preservation of wood. Under 'extraordinary constituents of plants' appeared references to colouring matters and their application in dyeing and printing, the preparation of salts of quinine and morphine, 'etc.'.

As an alternative to the general analytical course, the new Professor of Practical Chemistry, Alexander Williamson, offered a less intensive course to medical students.[21] To conform with the purposes of the Birkbeck Laboratory endowments, Williamson also provided a special evening course of fifteen two-hour sessions 'suited to the convenience of persons practically engaged in manufactures'. In fact, the content of these elementary courses for the medical students and the manufacturers was the same, stressing basic general chemistry taught by means of qualitative and quantitative analysis. Thus University College like the Royal College of Chemistry offered a single chemical syllabus to students with a wide diversity of career ambitions, while emphasising the special relevance of each course to the needs of its proclaimed audience.

The pattern of the Royal College of Chemistry and University College was repeated at King's College, where W. A. Miller, another German-trained chemist, replaced Daniell in 1845. During the 1850s several of the London medical schools hired new chemists as the first post-1815 generation of teachers died or retired. George Noad, an alumnus of the Royal College of Chemistry, had succeeded Brande at St George's in 1848. A decade later Frankland succeeded John Stenhouse at St Bartholomew's.[22] Many of these had been trained at the College of Chemistry. As we have seen, several other new posts were created in the 1840s, and most were taken by products of the new metropolitan colleges. They propagated the curriculum through which they had been taught. When the Royal Military Academy had to appoint a successor to Faraday in 1851, it specifically sought a man with research training.[23] The post was awarded to F. A. Abel, a Royal College of Chemistry alumnus and teaching assistant who had published a few competent analytical papers. He copied the model that he knew and valued.[24] Over thirty other teaching posts were occupied by graduates of just the first eight years of the Royal College of Chemistry. Through the challenge of its curriculum and the teaching of its students the pattern of the Royal College of Chemistry was widely disseminated.

The University of London science degree

By the late 1850s the combination of professional training and liberal education through chemical research had come to be well established. The educational utility of a core curriculum for science was the central feature of discussions about new science degrees for the University of London. The question of offering these degrees arose in 1858, when the university's charter expired. Hitherto there had only been Bachelor of Arts and Medicine degrees. A new Bachelor of Science degree was now established. The issues underlying this innovation were perhaps best captured in a memorial from a group of eminent scientists to the university.

... society now appreciating the value of their [the sciences'] fruits in alleviating the wants of man practically, regards the pursuit of these sciences as Professions and honours those who follow them. The Academic bodies on the other hand, continue to ignore Science as a separate Profession ... [they should recognise science] as a discipline and a calling, and should place it on the same footing with regard to Arts, as Medicine and Law.[25]

The group, which did not include any chemists, sought the recognition that the chemists had begun to achieve in the 1840s.[26] They managed to persuade the university to set up a committee to consider the question. It was decided to adopt a scheme aimed at reconciling a broad general training, the traditional hallmark of university education in England, with an emphasis on specialist subjects.[27]

Science subjects would be accommodated to the university ideal of 'liberal education through mental discipline'. A graduate so prepared would bring a developed character, rather than knowledge, to life's endeavours. In science this was to be achieved through the study of principles and of the details on which they were based across a range of disciplines regardless of the candidate's eventual specialism.

In the student's first, or matriculation, year of the science course, he studied Latin, Greek, mathematics, natural philosophy, modern history, inorganic chemistry, and French or German. During the second year he prepared for the first BSc examination by studying mathematics, natural philosophy, chemistry (including two days per week in the laboratory), anatomy and physiology, and French or German. In the third year, in preparation for the second BSc, he studied logic, mental and moral philosophy, mathematics, natural philosophy, organic chemistry (including laboratory practice) and geology and botany. Even in the final year the degree was not specialist. Notwithstanding the representation to the committee on the utility of science education there was no hint of any directly utilitarian training.[28] The liberal education aspect of the degree was far more evident than were the aspects of application. Despite the memorialists' original claim that a degree was needed to prepare young men for the profession of science, in the event the University of London degree was a professional qualification only in so far as it was based on university training. It lacked the flexibility to personal needs that the specialist chemical courses offered. No wonder that, up to 1870, only fifty students took the specialist honours in chemistry. Yet the degree soon gained nationwide influence. It was available to anyone who could pass the relevant examinations, wherever he had studied. The London BSc therefore came to define the horizons of the new provincial colleges.

Owens College

Until the 1850s the teaching of chemistry in England was dominated by London. There, despite the important developments of the 1840s, the majority of the students of the science were still destined for medicine. In the '50s and '60s a new provincial centre emerged. More closely associated with the manufacturing districts, Owens College, a private institution, was founded in Manchester in 1851.

The college opened with a double aim of offering general education in traditional university subjects and of serving local industrial needs. The by now established rhetoric of chemistry was peculiarly appropriate to the college's character.

In a locality where men's minds and exertions are mainly devoted to commercial pursuits, it seems particularly desirable to select as an instrument of mental training a subject, which, being general in its nature, and remote from the particular and daily occupations of the individual, may counteract their tendency to limit the application and, eventually, the power of applying the mental faculties.[29]

Chemistry was one of the foundation Chairs and Edward Frankland, who had trained in Germany, was its first incumbent. At the same time as it valued the mental training that chemistry would give, the college instructed Frankland to highlight practical utility.[30] In his inaugural lecture he dutifully echoed the founders' claims that chemistry would be useful mental discipline and industrial training. He explained that his method would be the one which, as he said, worked so well in the laboratories of London, Giessen and Marburg (where he had studied). He offered a lecture course in systematic chemistry, and a laboratory course in practical analysis; he also taught evening classes.

Despite the apt rhetoric, and reasonable facilities, Frankland's course was not an unqualified success. During his tenure from 1851 to 1857 an annual average of twenty students attended the lectures while some seventeen took advantage of the laboratory facilities each year.[31] It was the best attended and most popular subject at the college, but few students recognised its other than practical merits. Frankland was dissatisfied with this situation, particularly since many students needed elementary instruction. Furthermore, he felt thwarted in his own research. Writing to an eminent German colleague, he expressed his frustration in 1856, his penultimate year at Owens:

It must be in the highest degree satisfactory to you to be surrounded by such a number of students so many of whom are engaged in original research. Unfortunately this is far from being the case here and there is rarely any further desire for knowledge than the testing of 'Soda-ash and Bleaching Powder'.[32]

Within the established framework of chemistry teaching, Frankland tried hard to satisfy the boys who only wanted to learn how to test industrial products. The first examination paper does show that certain aspects of technical chemistry were dealt with, though often in a theoretical manner.[33] Seven out of fourteen questions were devoted to applied chemistry; they required a familiarity with the theory of practice. Question 6 was 'Describe in equations the changes and decompositions which take place in manufacturing oil of vitriol from sulphur and nitrate of potash'. Question 9 was 'What is the composition of bleaching powder? Explain in chemical language the processes of bleaching and of calico printing in the discharge style.' Other 'applied' questions used applied topics as a vehicle for teaching general theoretical points, as in question 7, 'Suppose a manufacturer of nitric acid is offered two cargoes at the same price per ton, one of nitrate of soda and one of nitrate of potash, which should he take for the purposes of his manufacture, explaining your reasons for the selection, from your knowledge of equivalents?' or question 14, 'Explain the general principles upon which the extraction of the metals from their ores depends'. Others required some practical problem-solving, on paper at any rate. 'How could you prove the presence or absence of potash, soda, lime, and iron, in a sample of manure?' (question 13).

When it came to the basis of chemistry, Frankland was uncompromising. Despite his assessment of the interests and abilities of his students, theoretical questions were couched in an abstract manner. Question 1 was 'What do you mean by the term chemical affinity, and what influences have the forces of heat, light and electricity, and cohesion upon chemical combination?' The second question asked 'By what three principal laws is the formation of chemical compounds governed?'

In the following year the general chemistry lecture course was actually called 'Chemistry in its applications to the arts etc.'; and in his third year of teaching, Frankland introduced separate examination papers on quantitative and on qualitative analysis.[34] In the former paper, only two questions out of ten dealt with applications. Question 6 was 'Describe the methods you would use in the estimation of the commercial value of soda-ash, potash, nitre, peroxide of manganese, and bleaching powder.' On the qualitative paper, three questions out of ten were on practical topics of local interest. Question 7 was especially pertinent: 'Suppose you had to examine the drainage water of a town with regard to its value as a fertilizing agent, what considerations would direct your researches, and how would you conduct your experiments?' In 1853-54, in an attempt to make his offering more attractive, Frankland instituted a separate course of 'technological chemistry', but it attracted only seven students.[35] On the evidence of the examination paper for 1854, the short course was general and descriptive, covering a wide range of industrial processes.[36] Despite its title, it is difficult to visualise the utility of such a course to practitioners in a particular area.

Frankland left Owens College for the more congenial metropolis in 1857, and was succeeded by the pupil of Alexander Williamson and Bunsen, H. E. Roscoe.[37] Roscoe represents the third generation of academically trained English chemists. A graduate of the University College laboratory, he took for granted the courses which his predecessors had had laboriously to establish. In his work and lectures can be seen the explicit rhetoric of pure and applied science. He was a successful researcher and became an eminent textbook writer as well as an industrial consultant.

Under Roscoe, chemistry began to flourish. He successfully translated the appeal that had been directed to the metropolitan establishment into a provincial context. Promoting his department, Roscoe threw himself into the affairs of the college. His main audience was a culturally rich Mancunian middle class. The practicality and moral worth of science were well recognised. Roscoe, appropriately, was to be the biographer of Dalton, hero of Mancunian science.[38] His German experience had also impressed him with the sanctity of the discipline, having its own values and problems. He expelled a student from his course for faking experimental observations.[39] He had a faith in the virtue of academic research. Roscoe could portray himself as the quintessential academic who built an eminent research school,

pursued research and won esteem.[40] Though at the same time he did
do a lot of consulting and became a director of the firm of
Castner-Kellner (pioneer manufacturers of caustic soda through
electrolysis), this was not central to his life. His voluminous
autobiography did not even mention consulting activities.[41] In fact
Roscoe could make a very adequate living from his academic work. It
was estimated in 1885 that his annual earnings from the university
amounted to £2,000, making him the best-paid professor.[42]

From Frankland, Roscoe inherited a small class in a dying
institution.[43] Within six years prosperity was firmly restored. In
1864-65, 100 students out of a college-wide enrolment of 128 took the
chemistry lectures while forty-nine attended the laboratory.[44] Roscoe
reorganised chemistry teaching at Owens into a three-year curriculum
of systematic study in conformity with the regulations for the
University of London's new BSc degree. However, at Owens as in
London, only a minority of students enrolled for the full course. It
was possible for the majority to study whatever subjects they wished
on an occasional basis.[45] As in London, the laboratory was open six
days a week and students could elect how long they wished to spend
there.

Roscoe's curriculum was outlined in the Owens *Calendar* for
1862-63.[46] There was a two-session lecture course consisting of fifty
lectures on inorganic chemistry (laws, properties, and preparation of
substances) followed by a course of fifty lectures on organic
chemistry. This could be followed by one lecture a week on
Technological Chemistry, which focused on 'the Chemical Principles
involved in the most important chemical manufactures'. Clearly, this
was not an especially significant part of the course. The student could
then proceed to the laboratory course. Roscoe was unusual in
suggesting a mimimum attendance. Although it was not mandatory,
he recommended that a student should attend for no less than four
days a week during three whole sessions. This would be desirable, for
the aim of the laboratory course was not unambitious.

The aim of this course is to make the Student practically acquainted with
Chemical Science, to enable him to conduct analysis and original research,
and to fit him for applying the science to the higher branches of Art,
Manufactures, and Agriculture.[47]

This was supplemented by an exercise class on chemical calculations.
Roscoe's examination papers for the early 1860s required a good deal

more calculation than had Frankland's, with more emphasis on theories regarding particular sets of phenomena than on general abstractions. If he did not really provide pupils with a knowledge of the practical details of chemical industry, he did give them a training in basic chemical skills.[48]

Roscoe wished to build a research school.[49] To do this, he had above all to attract students and collect local patronage. To the fathers of possible students he explained the integrity of chemistry; technology should be based on a firm understanding of the principles of science.[50] To his colleagues and pupils he extolled the educational utility of research. While mathematics with its rigour of deduction served the mind in one way, chemical research developed it in another.[51]

Roscoe shared many of the values of Hofmann; but he took their implications a stage further. He clearly differentiated between 'pure' and 'applied' chemistry. The former lay within a self-contained discipline, the latter was practised in industry by using the principles learnt in the university. In the 1840s phrases such as 'purely scientific' and 'science applied to the arts' had been familiar. 'Scientific' and 'practical' chemistry were the common antonyms.[52] The transition to the phraseology of pure and applied implied a relationship that would be demonstrated through Roscoe's courses. In Germany too Liebig was shifting his emphasis from the essential practical utility of chemistry to the notion of science as an end in itself.[53] At the same time as he valued the activity of research, Roscoe also ascribed educational value to the knowledge itself of 'pure science'. Though in practice, perhaps, the difference between his laboratory and that of the Royal College of Chemistry was not marked, Roscoe's rhetoric of the primacy of 'pure science' was characteristic of 1870.

With regard to chemistry, I should be inclined to say that the greatest and most important discoveries in the applications of the science to the arts and manufactures have not been made by men studying for the sake of the applications of science, but have been made by men who have been thoroughly instructed in the highest way in pure science.[54]

Pure chemistry clearly was to be the principal concern of the professor. Only when his teaching had been well instilled in the intending manufacturer could the pupil learn applications of chemistry. Such a division of labour was perhaps best expressed by Roscoe's old teacher, Williamson, in his 1870 inaugural lecture to the newly established Faculty of Science at University College. Entitled

'A Plea for Pure Science', the lecture argued for a division of labour in science.[55] On the one hand would be the professoriate who taught and advanced 'pure science' in universities, and, on the other, manufacturers outside, who, having studied, would practise 'applied science'.

Meanwhile education in chemistry was spreading. By 1863 there were twenty-one full-time professors of chemistry in London.[56] From the stimulus of the Royal College of Chemistry, teaching laboratories, involving unprecedented expense, were multiplying. By the 1870s the provision of a teaching laboratory for chemistry was considered essential by the many new provincial colleges.[57] Between 1850 and 1870 thousands of students passed through the leading schools. At Owens alone Roscoe boasted that he had taught 2,000 students by the time he retired in 1886.[58] His 'laboratory', built in the early 1870s, included a lecture theatre for 380 pupils, two main laboratories of over 2,000 sq. ft. each and more than twenty other rooms and laboratories encased in Waterhouse's spectacular neo-gothic building; it was a veritable cathedral of chemistry.[59] The science's ability to contribute to many industries and professions, to build the individual mind, and the virtue of the discipline itself had become well accepted.

Applied science: an alternative view

Roscoe's and Williamson's view was that 'applied science' was the application in industry of the theories of pure chemistry. This meant that the principal responsibility of academia was to create and transmit that pure science. An alternative view was held and strongly argued by Lyon Playfair. Citing Continental precedents, he argued that applied science itself could be an academic subject. This was not to depreciate abstract science, but to accord applied science a separate and equal academic status. Practical problems would be dealt with according to their industrial rather than their abstract interest.

Born in 1819 in St Andrew's, Playfair had studied under Thomas Graham first at the Andersonian and then as his assistant in University College London before completing his education at Giessen.[60] From the Scottish perspective that Playfair valued, his concern with applied science, informed, certainly, but not limited by pure or abstract science, was nothing novel. Playfair was unusually aware of his Scottish background. His inaugural lecture as professor at Edinburgh was devoted to the history of chemistry at the university.[61] He paid special attention to Black. When he replaced the accumulated obsolete apparatus in his laboratory, instead of allowing it to disappear, he safeguarded the legacy of Black by giving it to the newly established Industrial Museum.[62]

Moreover, Playfair spent fifteen years in England solving practical chemical problems for government and private interests. His role was similar to that of the many other consulting chemists in the metropolis. When the category of 'analytical chemist' was introduced into the London Post Office Directory in 1854 his was one of the first entries.[63] From the point of view of the metropolitan practical chemists, Playfair's division of pure and applied science was again unexceptional. His personal integration of Scottish academic and London professional traditions was characteristic of the 1840s. The Chemical Society had been founded in 1841 precisely by representatives of these two groups. It was appropriate that it was the young Lyon Playfair who in 1848 was asked to prepare the first draft of the Society's charter.[64]

Playfair was also well aware of what was going on abroad. He had studied in Germany and in 1852 he reported on the means of industrial training available on the Continent. One central concept clearly derived from that experience — his faith in the Continental polytechnic. In a speech on industrial instruction on the Continent Playfair showed how impressed he was by the Parisian Ecole Centrale des Arts et Manufactures, in which aspiring manufacturers were taught by eminent scientific men.[65] However, the conception of applied science through which he saw these institutions was home grown.

From his early twenties Lyon Playfair was politician as well as chemist. It was he who at the age of twenty-one intoduced Liebig's 'agricultural chemistry' to the British Association. This won him recognition and marked him out as a young man who would go places.[66] On the recommendation of Graham and Liebig he obtained a well paid first position as the chemist to a leading calico printer in north Lancashire.[67] Despite his initial delight with his laboratory and salary, the diminutive Playfair did not find sufficient stimulus in Clitheroe. His mind soon turned to academic advancement. He created what opportunities he could for himself locally, but his principal objective was London. The Geological Survey's ideas of creating a chemical side seemed to offer an attractive prospect of a potential job. He formally proposed to the government that he be allowed to teach agricultural chemistry at the Museum of Practical Geology with an emphasis on 'the theory of practice'.[68] His bid failed, though he did win the promise of government support, and for seven years he concentrated on a series of practical investigations, becoming in effect the government's consultant chemist, from 1845 as chemist to the Geological Survey.[69] Most notable among his projects was work for the Health of Towns Commission, an analysis of the waste heat from blast-furnaces conducted with Bunsen, and a study of coals for the navy. For the last he built a veritable factory to examine the suitability of various kinds of coals, employing two (and sometimes three) chemists and a technician.[70] In the year 1849/50, he also analysed food supplied to workhouses for the Poor Law Authorities and constructed dietary tables ('although the enquiry does not come within the terms of our Museum', as he noted in the annual report); ensured the purity of metals from which guns were made for the Ordnance Department; analysed water for the General Board of Health; analysed recent and fossil shells in connection with marbles and limestone; examined a new class of salts possibly useful for dyeing (this produced his only paper to the Royal Society) and coals for gas-making.[71] Playfair was also engaged on agricultural issues, as chemist to the Royal Agricultural Society and member of the commission on the potato blight. He carried out too his first, and last, ambitious piece of specialty-oriented research: he worked with Joule in an attempt to establish a theory of solutions, but the results were disappointing.[72]

Playfair seems to have been encouraged to form a school of science. An agreement with the government promised him the position of chemist to the Geological Museum on the demise of the aged incumbent if by that time Playfair should 'have succeeded in organising a body of competent pupil Assistants at the Museum'.[73] By 1850 all that could be said was that two assistants had been provided, of whom one was a pupil. In 1851 a new improved building was constructed in Jermyn Street and the museum was upgraded to be the new Government School of Mines and Science Applied to the Arts. Before it opened the Professor of Chemistry died, and on the grounds that his one student plus one other assistant constituted having organised 'a body of competent pupil Assistants' according to the original agreement, Playfair inherited unequivocally the position of chemist to the new school. His main responsibility was 'Chemistry in its application to Agriculture and the Arts'.[74] By the end of February

1852, in the first academic year, Playfair had nineteen students and two assistants.[75] Although industrially oriented, his course dealt narrowly with analyses. A typical examination question was:

A salt containing potash, lime, sulphuric acid and water, being analysed gave the following results: – 10.05 grs. of the salt gave 3.05 grs. carbonate of lime and 14.85 grs. chloride of platinum and potassium. 9.28 grs. of the salt gave 0.51 grs. of water and 13.22 grs. sulphate of barytes. What is the composition of the salt in 100 parts, and what is its probable formula?[76]

The significance of Playfair's attitude to chemistry at the time, however, was not principally transmitted though his own teaching. A self-confessed 'lover of adulation',[77] Playfair's big break had come in 1850 when he was asked to step in at the last moment to help organise the Great Exhibition of the following year. Through this work he became friendly with the Prince Consort and involved with the considerable problem of what to do with the £180,000 profit from the Great Exhibition.[78]

Playfair felt with his patron, Prince Albert, that the most appropriate use of the exhibition profits lay in the conversion of the School of Mines into a Continental-style polytechnic. In the autumn of 1851 he put the suggestion to the Director, Henry De la Beche, who was to come out vehemently against the scheme to generalise his institute.[79] In November Playfair delivered a major address, 'The Study of Abstract Science essential to the Progress of Industry', as his introductory lecture given to the School of Mines.[80] Another forum for his arguments became available the following January. On Prince Albert's recommendation the Society of Arts held a series of lectures on the results of the exhibition. Playfair made perhaps the most eloquent of the speeches. In his lecture 'The Chemical Principles involved in the Manufactures of the Exhibition' he argued that industry should become a profession and training for it should be in a technical university.[81] Science would be at the centre of the curriculum. In this way the status of both science and industry would be raised. He followed up this speech by a tour of the Continent, and his introductory lecture to the School of Mines the following year was a closely documented statement of the progress of industrial education on the Continent.[82] Soon, he threatened, Britain's industry as well as her education would start falling behind. Playfair was echoing messages that were coming to the government directly from industry and from the Society of Arts. In November 1852 the Queen's

Speech announced that the government would devote the proceeds from the exhibition to technical education, and the following March the government agreed to form a new Department of Science and Art within the Board of Trade. Playfair became the Secretary for Science.

Playfair's new status gave him influence, and he saw it as a way of promoting his view of a national system of science education. With Henry Cole, his partner in the new department, he wished to establish a regional network of science schools with a great central organisation in London to train the teachers: the newly retitled Metropolitan School of Science Applied to Mining and the Arts (for which the Department was also responsible).[83] There were already well established Schools of Design run through the arts section of the department under Cole's supervision. Playfair envisaged the integration of the two systems into schools of industry. He wrote to Henry Cole:

In plainer language your colours as well as your glazes are chemical appliances requiring as much practical skill as knowledge, that men of abstract science are insufficient to improve them greatly, but practical men acquainted with Science would exercise a more beneficial influence on them. Why should you not have a School of Industrial Knowledge instead of Art only. Birmingham is already doing so, Manchester is discussing such a project, and the signs of the times point to it as a necessity.[84]

De la Beche fought against the conversion of his beloved School of Mines into a teacher training college for all industry. He resisted Playfair bitterly, but his death in 1855 almost gave Playfair the opportunity he had been looking for to separate the Geological Survey, the Museum of Economic Geology and the School. Unfortunately for Playfair's scheme, the predominantly geological professoriate preferred De la Beche's vision and rebelled. The aristocratic geologist Roderick Murchison was invited to take the position as Director. He had clearly opposite views to those of Playfair. He saw the institution 'simply as a School of British Geology and Mines. The affiliated Sciences are all subordinate to that fundamental point.'[85] Murchison's accession for the time being prevented the redirection of the School of Mines.[86] The narrowness of Playfair's defeat is indicated by a prospectus of the school dated July 1853. As it was first drafted it reflects Playfair's views but it was amended to what became the published version.

The Autumn Session will commence on 1st October when courses of lectures and practical demonstration in Science will be given in the Metropolitan School of Applied Science [the 'Applied' was crossed out] in Jermyn Street and in Art at Marlborough House ... The Courses are intended to impart a Knowledge of the principles of Science & Art involved in Manufacturing and Mining processes ['Mining' in margin, added later] to those who may desire to carry them into practical and industrial pursuits. Special attention is devoted to the training of teachers in a knowledge of Science and Art.[87]

Playfair was disillusioned; he had lost. The other side of his plan, the regional science schools, were hardly more successful.[88] His view of the ideal seems to have been the Watt School of Arts in Edinburgh.[89] Founded in 1824 on the model of the Andersonian Institution in Glasgow, the school gave evening classes in science to working men. The Watt School was given a £50 scholarship to send its best pupil to the School of Mines in London. Some regional science schools were actually established in the early 1850s. In Bristol a Trade School which would be often referred to, in Wigan a mining school, in Newcastle and Aberdeen science schools, and in Birmingham the 'Birmingham and Midland Institute'.[90] Playfair went round the country propagandising. In Sheffield in 1853 he gave prizes at the Sheffield People's College, a mechanics' institute. His speech, entitled 'Science in its relation to labour', urged the locals to convert their college into a science school. 'Science is a religion,' he urged, 'and its philosophers are the priests of nature.'[91] Though a class on the applications of chemistry was held the following year in Sheffield, Playfair's appeal for a major reorientation was unsuccessful.[92]

Despite Playfair's untiring propaganda few schools were established and fewer succeeded. In 1858 he suggested to the Science and Art Department that the real reason for the successful spread of Schools of Design was that they had been able to rely on wealthy patrons to pay for expensive daytime classes for their children and thereby subsidise cheap evening classes. Science did not seem to be able to attract the same kind of support.[93]

For a few more years Playfair held a post at the department but in 1858 he took the newly vacated Chair of chemistry at Edinburgh. He spent £1,000 of his own money on fitting up the laboratory, and got the university to expend £500 on furnaces, fume cupboards and distillation equipment.[94] He lectured to about 200 pupils, mostly medical, had a practical class of seventy-two and about a dozen in the more advanced analytical laboratory.[95] Despite his faith in the teaching of applied science, his course seems to have been conventionally academic.[96]

However, within a few years Playfair was back to politics. In January 1863 he proposed to the Senate of the University of Edinburgh the establishment of degrees in science, and in 1865/66 chemistry was for the first time accepted as an honours specialisation within the Faculty of Arts.[97] Yet more ambitiously, Playfair sought to set up a new centre of 'Applied Science'. This took him beyond chemistry. He sought funding for new Chairs in agriculture and engineering from the Highland Society, the government and from a wealthy industrialist.[98]

Playfair and the Professor of Agriculture, John Wilson, who had once been his assistant in London, drew up a memorial to the Treasury signed by the Principal of the university, Sir David Brewster, who had himself long been an advocate of applied science. The appeal was successful. The Treasury matched private endowments with £200 for the engineering professor and £150 for agriculture.[99] In April 1868 the Secretary of the Senate tried to hasten formal approval, pointing out that the university's action 'at a time when the attention of the nation is specially directed to "Applied Science" is likely to be of great advantage to the University'.[100] Though objections were later raised, the prospectus for 1868/69 carried advertisements for degrees in 'pure' and in 'applied' science.[101] The former included 'Physical and Natural Sciences: Natural Philosophy, Chemistry, Zoology and Geology, Botany, Animal Physiology'. The latter comprised 'Agriculture, Engineering and Mechanical Science, Veterinary Surgery'. It may be striking, given Playfair's background, that chemistry was counted as a pure rather than as an applied science. He would have liked to have had courses in mining and for future managers of chemical works but there were not the resources.[102] When asked whether he had not already taught applied science for some time, he replied: 'No, not methodically; but we have illustrated our lectures in pure science by industrial applications.'[103] While in contrast to other academics he argued that in principle chemistry itself was not the basis of industrial practice, he had not resolved the outstanding problem of a pedagogy of applied chemistry. Nor did he pursue it. In 1868 Playfair was elected to Parliament and abandoned for good the narrower world of chemistry.

Playfair's goal had not been principally chemistry at all but rather applied science. He lacked the specialty orientation that was increasingly characteristic of the new professoriate. The laboratory that he had proposed in the '40s was very different from

Hofmann's.[104] As a contributor to the discipline itself he was not highly regarded. In 1865 Henry Watts, editor of the *Journal of the Chemical Society*, fearing that Playfair would attempt to succeed Hofmann at the RCC, wrote to Roscoe:

Hofmann is going to Berlin & it appears to be quite settled; but I have not heard *when* he will depart from among us. There is a good deal of talk as to who is to fill his place. Playfair intends to go in for it and seems to make cock-sure of winning. This is generally thought to be too bad — and accordingly some people are endeavouring to stir up Williamson to put himself forward as a candidate: but he appears to be afraid of being beaten by Playfair which of course would not be pleasant. Frankland seems to be quite satisfied with what he has got. But surely someone ought to be found plucky enough to take the bull by the horns. What do you think of trying your fortunes? As far as scientific qualifications go, everybody would of course place you miles above Playfair. The only question would be about influence. If Playfair were to get the place it would of course be because he has friends at court — and though not really a great chemist, has somehow managed to persuade the outside world that he is so. But I think your position is good enough to enable you to fight P. even on his own dunghill.[105]

Playfair's investigations had brought him a reputation with the consumers of science, including the government. They had little relationship to the pressing problems of the discipline as an intellectually coherent enterprise. As a professor he was unsuccessful; his peers had little respect for him. Roscoe noted in an obituary, 'If Playfair had remained under the influence of Dalton and Joule, his record of original work would probably have been much longer than it is, but his activity was destined to be turned into other channels.'[106]

Mass education

The belief of an increasing number of academics in the possibility, but not in the primacy, of training in applied science jarred with the original vision of Playfair. After all, preparation for industry was his central objective. The foundation of university Chairs in engineering testifies that Playfair's philosophy did articulate an enduring theme in British culture.[107] However, for the moment, failure, for administrative reasons, of Playfair's scheme for science schools under the Department of Science and Art also meant that his vision lost influence on the government. Soon after his departure from the department, a new scheme which was to become very successful was started. In 1859 the famous payment-by-results system was instituted. Emphasis was shifted from integrated science schools to evening science classes leading to examinations. Teachers were paid according to the number of examination successes achieved by their artisan pupils. The system laid great stress on the centralised examination curricula. Initially the subjects were: Practical and Descriptive Geometry, Physics, Chemistry, Geology and Mineralogy (applied to mining) and Natural History.[108] The curricula were designed by the metropolitan scientific elite. In the case of chemistry, this meant the professors at the Royal School of Mines – first Hofmann and then Frankland. Conspicuously Playfair was not at all involved. It is hardly surprising therefore that the style of the curriculum in chemistry closely conformed to that of university-level teaching.

The emphasis of the chemical curriculum was on general theoretical chemistry buttressed by factual knowledge of properties and preparation of some major compounds of the principal elements. Students were expected to have a knowledge of practical chemistry and they received the usual smattering of descriptive knowledge about a wide range of industrial processes. There was no practical examination until 1878. It has been suggested that the standard of the advanced paper in inorganic chemistry could be compared to that of modern British school examinations.[109] That this comparison can be made highlights the academic nature of a curriculum that was meant to help transform industry. Exams were set and marked by academics. The ethos was stressed by a London BSc in chemistry who had joined the Birmingham and Midland Institute in 1863. In a forty years' retrospect he declared:

Be assured, the true technical education is that which grafts on to the sturdy trunk of the practical man the habit of thought of the scientific worker. Work steadily, without thinking of immediate reward, and my impression is that the bread cast upon the waters will return after many days.[110]

Characteristically both Hofmann and Frankland commented with pleasure on the fact that teachers seemed to be keeping abreast of the latest chemical theories, since the new chemical notation of the middle '60s was clearly appearing in examination papers.[111] At the same time, however, Frankland noted weaknesses in areas such as calculations and noted that answers to the analytical questions seemed to be based on textbooks rather than on practical work in a laboratory. In this he was probably quite right. Few students would have had access to anything more than demonstrations.[112]

When the funding programme for the Department of Science and Art examinations began, chemistry was by any measure the dominant subject. It accounted for over a third of the papers sat. The subsequent growth of the system was indicated by the all-important examination tally. From 1860 to 1870 the numbers taking the inorganic chemistry paper increased more than eightfold, from 382 to 2,694.[113] Even so, the growth of chemistry was not unusually fast. In the 1860s the total number of papers taken increased more than four times more rapidly. Even though chemistry was important, it would seem that its syllabus was not overwhelmingly attractive.

Towards the end of our period chemistry, though still growing phenomenally, began to be overtaken by other subjects, arguably subjects which were more pertinent to industry at the time. Thus in 1867 chemistry, with 2,311 students, lost its lead and declined to third place behind practical, plane and solid geometry and machine construction and drawing, with building construction and naval architecture not far behind. By 1870 chemistry was in fifth place. Its head start as the archetypal science had been lost.

The ambiguities of the Science and Art Department course reflected the problem of curricula as a whole. Chemistry was justified because it was useful. Yet it was questionable how useful in fact a summary of the science's multiple applications could be. On the other hand the treatment of the fundamental principles might prove a useful background in future life even though it was not obviously practical. But again, as we shall see, the realities of industry and the chemical community did not suggest that pure chemistry was daily being applied to great inventions.

Chapter Four:
Chemistry in practice

One could not of course discover a 'reality' with which to compare the rhetoric of the professors. However, one can investigate chemists' careers separately from the academics' projections of them. The effort is intended not to prove the promotional literature 'wrong', but to help observers more than a century later understand its import. For in practice theoretical science was much further from practical application than might be inferred from a first reading of the professors' words. As will be seen in the next chapter, this problem came to entail critical intellectual and institutional questions.

Students

Despite the growing interest in other subjects, until 1867 chemistry attracted the largest number of Science and Art Department examinees. What did this mean? The system had been established to help industrial workers gain an understanding of the scientific basis of their trades. However, an 1863 survey indicated that in practice artisans constituted barely half the student body.[1] Though by 1871 in some industrial areas the proportion had risen to 75%, application in the factory was no more an adequate description of Science and Art Department student motivations than of others' interests in science.[2]

The chemical students can be more specifically characterised from the evidence about the sixty-five students who became medallists in chemistry to 1870.[3] Though it must be recognised that these men who had achieved a high standard in their exams may not have been typical, there were few medallists with industrial inclinations, and only one person styled himself a manufacturing chemist. This is particularly striking in view of the concentration of Science and Art Department teaching in the north of England. Although twenty-one different occupations were represented, the most numerous were fourteen teachers, followed by ten 'students' and six people who styled themselves chemist's or laboratory assistant, four who were chemists, three druggists and one analytical chemist. There were five clerks and two engineering apprentices. On the whole, chemistry, like other Science and Art subjects, was not serving as the scientific basis of artisans' pursuits.

In the 1870s Frankland classified the audience for chemical tuition as medical students, artisans and men destined for technical pursuits.[4] About most of these pupils and indeed most of the schools we know little. However, records of some of the chief colleges survive in some detail. Henry Roscoe at Owens College obligingly conducted a survey of the career goals of his laboratory students in 1870.[5]

A small minority of the alumni of the colleges entered academic life. Men trained at Owens and the Royal College of Chemistry staffed many of the new laboratories of the late nineteenth century. Men such as Bedson at Newcastle and Kipping at Nottingham carried on the academic tradition and in turn trained academics influential in the twentieth century. However, future academics accounted for only a small minority of the students. Both Owens and the Royal College of Chemistry taught fully matriculated students engaged in three-year courses. These included candidates for the comprehensive London BSc degree, established in 1858. Again such students were relatively few. Far more numerous (twice as many at the Royal College of Chemistry) were 'occasional' students who only took chemistry courses.[6] Most of the occasional pupils, even at the elite colleges, clearly did not stay very long. At Owens College, two-thirds of the entrants only stayed a year.[7] At the Royal School of Mines the figure indicated a similar pattern of attendance; out of the 469 occasional students who passed through the College of Chemistry in the eighteen years from 1853 (when the college had been incorporated in the School of Mines) only 20% (ninety-four) stayed more than a year. For most of the matriculated students, chemistry was but a subsidiary part of their course. Whatever the outcome of such an education, it was not a new identity as 'chemist'. Of the 706 alumni, both matriculated and occasional, of the Royal College of Chemistry laboratory between 1853 and 1870, only 13% (ninety-two) had joined the Chemical Society by 1870. If we look at publications, only 143 had published in any subject by 1883, and that number includes half (forty-eight) of the Chemical Society members. Almost a third of the occasional students who published came from the minority that had stayed at the college more than a year. By contrast, of the 158 occasional students who stayed one term, only eighteen published and twelve joined the Chemical Society.

We know in more detail about the actual career of some 60% of the Royal College of Chemistry occasional students who stayed more than a term. As Frankland himself noted from a less comprehensive list, most alumni of the Royal College of Chemistry tended to go into some branch of manufacturing.[8] Perhaps surprisingly, so far as can be determined, the Royal College of Chemistry students did not gravitate towards the chemical industries, although the industries they entered used chemical processes.[9] There were a considerable number of brewers, particularly from the Burton breweries. The coal tar dye industry and the leading London pharmaceutical firms were also destinations of student careers. However, I. L. Bell, the Newcastle ironmaster and chemical manufacturer, claimed to the Devonshire Commission that he had never met a Royal College of Chemistry graduate in a local works.[10] Henry Bessemer had attended the college though he does not mention it in his autobiography.[11] From Tyneside a number of less well known chemists had attended the college and were at least acquainted with Bell. However, his testimony does indicate that the previous thirty years of development had not had a dramatic impact on a major industrial area. At the same time it is true that there had been a shift in attitudes. In 1875 John Morrison articulated before the Newcastle Chemical Society the by then conventional view:

In fact, many of us have gradually taught ourselves to look upon the laboratory almost in the light of a necessary nuisance, an institution only of use to us to ascertain if we get full value in our raw materials, and that our finished products are sent out up to strength. Henceforward, we have said to ourselves, our chemistry must be confined to productions and costs per ton, and, in Lancashire especially, to quantity turned out per furnace; the best manager being he who, *prima facie*, can get the greatest number of tons out of a given number of men and a given amount of plant.[12]

In Manchester, too, this modest role was propounded by Roscoe. He envisaged that his own pupils would primarily become analysts and production managers.[13]

So most alumni of the schools had just passed through briefly. Only a minority of ex-chemical students retained their connections with the academic community. Such were the few attracted to the artificial dye industry. Hofmann's research programme was deeply concerned with the chemistry of coal tar and its products. One of his pupils was the dedicated W. H. Perkin, who discovered the first coal tar dye, mauve, during his four terms at the college. He immediately entered manufacturing and later discovered a synthesis of the industrially important dye, alizarine. However, in 1874 Perkin faced the need to make a renewed commitment to manufacturing if he were to compete in the alizarine industry. As a long-established member of the Chemical Society he decided to leave manufacturing and to devote himself to science. His career reflected the dual orientation of the academically inclined students who entered manufacturing. Though few in proportion to the whole, they were nevertheless quite significant in their numbers and they were so concentrated in a few industrial sectors that in those areas they were conspicuous. These men provided the links with the academic elite for others whom they influenced and taught. They were the members who made up the manufacturing population in the Chemical Society. Thereby the original coalition of diverse chemical interests represented in the Society was subtly changed. Increasingly those of the new generation of manufacturers who did join had academic links. The membership though still diverse became increasingly dominated by academic chemistry.

Chemical Society and the chemical community

Between 1841 and 1870 the membership of the Chemical Society increased from seventy-seven to 554.[14] By the end of the period 944 associates and fellows had been elected. At the same time the Society was firmly ruled by an oligarchy slowly developing from the band of founders. Council was a formal affair whose meetings confirmed private deliberations. Power was held by special sub-committees such as the publications committee and by the officers. In the first few years important offices were held by the oldest metropolitan professional chemists who had been influential when the Chemical Society was founded. From 1852, however, with rare exceptions, academics held all the reins of power and influence. The academic elite of the Chemical Society encouraged recruitment. In the early years Graham or Robert Warington signed 90% of the application forms.[15] The forms dating from 1852 to 1865 have been lost but those from the later 1860s again indicate the importance of an inner circle.

From the beginning the bulk of the members seem to have been academics, manufacturers and consultants. In contrast with the conventional view of nineteenth-century British science, amateurs were rare and those who joined were exceptional men. Physicians and pharmacists did join but in insignificant numbers. The leadership was successful in maintaining a rough numerical balance among the three dominant categories.

At a rhetorical level one finds presidents returning to the theme of the unity of science and practice. In 1857 W. A. Miller, professor at King's College London, in his presidential address could announce Perkin's discovery of mauve as a proof that chemistry could be both academic and useful. 'One of our Fellows, Mr Perkin, has afforded me the opportunity of bringing before you the results of a successful application of abstract science to an immediate practical purpose ... .'[16] Of course this was not literally true, since Perkin's discovery had been an accident, alighted upon in an attempt to achieve a different practical result, the synthesis of quinine. However, more important than this historical quibble was the implied moral that academic research could relate to manufacturing. A few years earlier, in his presidential address of 1853, Daubeny had spoken of Professor Bunsen's work on volcanic activity as 'undeniable evidence of the extensive utility of our pursuits'.[17]

The rhetoric served to express a unity based more on the strength of particular networks linking academics and industry than on the grand interplay of science and practice. While it is useful to use categories such as manufacturer, professional and academic, the relative importance and meaning of these categories varied from region to region. About half the members lived in London and its environs. Of the remainder, the majority worked in the industrial towns. Though the great alkali districts were poor in Chemical Society members compared to London, they were strongholds in comparison to the south, west or east of the country. Out of 859 with known addresses in Britain, 446 members joined from London, 104 from the north-west, principally Lancashire, and seventy-six from the north-east. A further sixty-nine joined from the region of Glasgow and from the Midlands.

The industrial regions had their own characteristic forms of chemical institutions. With the exception of Glasgow, none had old-established universities. However, this paucity should not blind us to the real institutional wealth of the industrial provinces. From the 1840s private schools had come to offer tuition to the masses and assistantships to the few. As earlier in London, professional chemists could make a living through a combination of roles. In Manchester Lyon Playfair founded a teaching laboratory in the basement of the

Royal Institution, where he was an honorary professor.[18] He used the laboratory for carrying out remunerative analyses for the Royal Agricultural Society. When Playfair left for London in 1842 he was soon succeeded by Frederick Crace-Calvert, who had learnt his chemistry from a leading industrial savant in France. He used the laboratory to make carbolic acid, popularised by Lister.[19] The need for assistants to help both Playfair and Calvert gave work to well educated Chemical Society members. Playfair was assisted by the Giessen-trained Robert Angus Smith, later the first alkali inspector.[20] Crace-Calvert employed several foreign assistants, including Zurich-trained Henry Brunner, whose brother John was the partner of Ludwig Mond.[21] At the lower middle-class equivalent to the Royal Institution, the Mechanics' Institution, there was also a laboratory. Daniel Stone, the Edinburgh-trained lecturer in chemistry who worked at the Manchester School of Medicine, opened a teaching laboratory there during the 1840s.[22] In the next generation Stone's pupil Fernside Hudson taught government science classes from his consultant's laboratory in Corporation Street.[23] Trafford College in South Manchester was administered by another Chemical Society professional chemist, James Hudson, organiser and chronicler of the mechanics' institutes.[24]

When the young professional chemist William Crookes, who had been trained at the RCC, considered moving to Manchester in 1864, he consulted Robert Angus Smith, who was already established there. The advice was 'don't'. It was impossible to make a career as an analytical chemist in the town, though it was true that about five men had been successful as consultants. Crookes was properly rather sceptical and wrote to his friend Peter Spence enquiring whether there was 'any mental or constitutional peculiarity in Dr Angus Smith which inclines him to be extra cautious and desponding'. In fact Crookes was rather attracted by the off-putting picture given by Smith. There seemed to be plenty of little analyses that Smith had refused to do. Potential pupils appeared to be plentiful and he wrote to Spence, 'there are also to be considered the chances of profit from entering more into manufacturing or dabbling in patents'.[25] Only the ultimate decision of Crace-Calvert not to sell his laboratory prevented Crookes's move to the provinces.

The Manchester pattern was repeated with local tones in many major northern towns. In Glasgow the professor of chemistry at the Andersonian Institution, Frederick Penny, earned £6,000 a year from consultancy.[26] In Liverpool the ambitious student of Liebig and scion of the alkali-making Muspratt family, James Sheridan Muspratt, founded a college of chemistry in 1848.[27] From 1854 the Birmingham and Midland Institute offered a laboratory and teaching to enthusiatic evening students. Among them was a reticent bomb-making would-be assassin of Napoleon III.[28] In Sheffield a college of popular science was taught by a local consulting chemist, A. H. Allen. In Newcastle the College of Medicine contracted Thomas Richardson, Liebig's first British PhD, local consultant and pioneering superphosphate manufacturer, to spend his little remaining time teaching its students from 1844. On Richardson's death in 1866 he was succeeded by his assistant, A. F. Marreco who, more influenced by models of the academic role, relinquished his consultancies to concentrate on the new position.[29] Other colleges where chemistry was taught to working-class or lower middle-class audiences could be found in Halifax, Leeds and Chester.[30] The teachers, professional chemists on the whole, tended to join the Chemical Society. By 1870 the Society had attracted almost all this burgeoning class. These men were not great researchers. However, they did teach large numbers and kept in close touch with industry.

Despite the Chemical Society's rhetoric, and the membership figures, manufacturing members came from relatively few companies, were unusually well educated and had uncommonly close connections with the chemical elite. Forty per cent of identified manufacturers had some chemical education. This situation was one step removed from that in the earlier part of the century. Then the relationship between chemistry and industrialists had been through the general cultural role of science. By mid-century the culture of chemistry itself linked provincial industrialists to the organised discipline. No longer local culture locally nourished, chemistry was cosmopolitan culture, nourished by national institutions. With the consultants, manufacturers shared a culture that valued chemistry. The difference between the 'academics' and the 'practitioners' who joined the Society was a matter of emphasis rather than a deep dichotomy. For example, the large Westmorland gunpowder industry was only represented in the society by two members, one of whom had been a pupil of Playfair at Edinburgh. They both came from the one company that strove to resolve scientifically the problems caused by the high price of saltpetre. While other manufacturers complained, the Gatebeck mill synthesised the needed potassium nitrate from readily available potash and sodium nitrate.[31] They were among the first in Europe to do so. The unusual manufacturers who did join the Chemical Society did not necessarily 'use' chemistry more than others. Rather they tended to have social links with leading chemists. The mere twenty members elected to the Chemical Society from Tyneside included the Glasgow and Giessen-educated Thomas Richardson and his assistant, A. F. Marreco, the now prosperous H. L. Pattinson, his pupil, John Pattinson, his son-in-law, I. L. Bell and close associates, Calvert Clapham and Georg Lunge who would one day be a professor at the famous polytechnic in Zurich. There were of course some who were not so well connected. The metal industry in Birmingham, Runcorn and Sheffield was represented in the Chemical Society by a few members with less close affiliations to Continental chemistry.

The largest regional concentration of manufacturing members was in the north-west. The calico printing industry had had an efflorescence of interest in chemistry during the 1840s. Most of the major companies employed at least one trained chemist. There were informal networks of people with chemical interests. We know of a group of nine men concerned with calico printing who met fortnightly to discuss chemistry in a pub in Whalley, north of Manchester.[32] Five of the nine joined the Chemical Society. Several works were very well represented in the Society.[33] Lyon Playfair's first employer, the leading calico printer James Thomson of Primrose, joined with his two sons. So did Benjamin Cooper, Playfair's successor at Primrose. At the Mayfield works in Manchester, not only the proprietor William Neild joined, so did his two sons and two employees – James Graham and John Thom, both pupils of James's brother, Thomas Graham. While the employees were well connected with the Chemical Society leaders through their education, the employers were active patrons. William Neild contributed to the building of the Birkbeck laboratory at University College London, while James Thomson was an important supporter of the Royal College of Chemistry.[34]

The dye manufacturers of the 1840s were concerned with natural and inorganic dyes. In the 1850s and 1860s emerged a new generation of organic dye manufacturers. These were particularly well educated, closely connected to the academic centre and well represented in the Chemical Society. John Dale and his son, both members, were pioneers of the new technology and employers of several of the German chemists who worked in Britain during the 1860s. Caro and

Martius, who were later founders of the great German dyestuffs industry both worked for Dale.[35] In Lancaster, Joseph Storey, a past pupil of Edward Frankland at Owens, set up in business in 1860 to make picric acid and pigments.[36] His first manager, A. Bickerdike, joined the Chemical Society and a few years later he hired another Chemical Society member. The soda industry was well represented in the Chemical Society from the beginning. Several of the Muspratts joined, as did other great manufacturers and innovators such as David Gamble, William Gossage, Henry Deacon and Henry Brunner. From alum makers Peter Spence came Spence himself, his relative, J. Berger Spence, and their chemist J. C. Bell.

Outside London and the great alkali and dye-producing provinces the participation of manufacturers in the Chemical Society was thin – only forty-three identified manufacturers were elected in the years to 1870. Most of those other provincial members whom we have managed to identify were academics. There were about twenty schoolmasters teaching at such public schools as Wellington and Hurstpierpoint, which were known for excellence in science, also surprisingly many Oxford dons, about twenty of them.[37] In contrast to the professionals in the industrial towns, these teachers at public schools and Oxford were only weakly connected with industry.

London was always the great exception. Although it did not have the specialised chemical industry of the north, it was a national centre for industries using chemical processes, including the manufacture of gas, soap, explosives, artificial fertilisers and the brewing of beer. Several of these were trades that could be found throughout the country. In London there were unusual numbers and good opportunities for links with academics. Among the founders of the Chemical Society, Robert Warington, the first Secretary, had worked for Truman's the brewers after a training under J. T. Cooper and then at University College. George Lowe, engineer to the Gas Light & Coke Company, was also a founder member and was a close associate of Cooper. Though these men were not typical they brought in a few of their colleagues. Altogether three men joined from Truman's. Only one other member came from any of the other London breweries. Similarly several members came from the Gas Light & Coke Company. Three rather different industries were well represented in London: fertiliser manufacture, artificial dyestuffs and government manufactures (particularly the Royal Mint and Woolwich Arsenal). Each was very modern and closely connected to the Royal College of

Chemistry. It is perhaps rather surprising that six men employed in the manufacture of artificial manures in London joined the Chemical Society. However, the industry had begun with Liebig's suggestion about the value of coprolites and was associated with well trained chemists. J. B. Lawes, who began the manufacture of super-phosphate, had studied at Oxford under Daubeny. He joined, as did several of his assistants. Benjamin Newlands, who studied at the Royal College of Chemistry, held several assistantships before joining Gibb's manure works on Victoria Docks. The synthetic dye industry had begun in London with Perkin's works at Greenford Green. To that was soon added a plant belonging to three other Royal College of Chemistry graduates, Simpson, Maule and Nicholson. The manufacture of synthetic dyes was perhaps the most 'chemistry'-intensive of all the chemical industries.

Government was a leading sponsor of manfacturing and analytical chemistry in the region of the metropolis. F. A. Abel one of the first generation of Hofmann students, went to Woolwich in succession to Faraday, just at the time that explosives were being revolutionised. By 1870 three Royal College of Chemistry graduates at Woolwich had joined the Chemical Society and in addition another chemist of uncertain education. The Admiralty chemist, William Weston, was also a Royal College of Chemistry man. The Mint employed trained chemists who joined the Chemical Society. So did the Inland Revenue laboratory in Somerset House, though that was concerned with analysis rather than manufacture.

Outside manufacturing there were the numerous analysts and consultants. Though some had dubious credentials, others were well trained. Men like Playfair developed the model of the professional chemist established since the the 1820s. Some moved between independence and industry – they included Edward Riley, who, having studied chemistry at the Royal College of Chemistry, moved to Ebbw Vale Ironworks for some years. He returned to London as a consultant and was a close friend and advisor of Henry Bessemer in the development of the steel rail. Others, such as J. T. Way and Augustus Voelcker, moved between academe and consultancy. Both were professors at the Royal Agricultural College in Cirencester before starting practices in London. Unlike their northern counterparts these consultants faced considerable competition from major academic institutions. The Royal College of Chemistry-certificated William Crookes initially found consultancy life difficult in London. He complained:

The public all go in one or two grooves – their work is sent as a matter of course to the College of Chemistry, or to some of the well-known chemical schools attached to hospitals. There is plenty of work to be done, but there are so many eminent professors twice my age and standing that they absorb it all.[38]

The concentration of academics in the metropolis was unique in Great Britain. Half the academics who joined the Chemical Society were from London. Altogether eighty-four academics from London were elected in the period 1850 to 1870. Their publications dominated the research output of British chemistry. More than 1,200 papers had been published by them by 1883. Three-quarters of them had published at least one academic paper by that time and more than a third had published more than ten papers.

Their research dominance gave the academics control over the content of the discipline, complementing their institutional power. Within a few years during the mid-1850s, chemists in London made major advances in understanding the structure of organic compounds, and discovered the first synthetic dye.[39] The biography of August Kekulé testifies to the intellectual excitement of the time as the chemical structure of matter was being elucidated.[40] At the same time, the growth of analytical training created an increasing number of opportunities for teaching and research assistants. Indeed, in the mid-1850s a remarkable academic community grew up in London, attracting many young German assistants and students, most of whom joined the Chemical Society. Five assistants of John Stenhouse at St Bartholomew's Hospital joined during the year 1854/55. Kekulé, a young graduate of Liebig's laboratory, for want of better prospects answered a call from John Stenhouse for a research assistant, and came to London in 1853. Another assistant was hired at the same time, Heinrich Buff, the son of an eminent professor at Giessen. Kekulé got to know Williamson, at University College, William Odling and Edward Frankland. They provided the stimulus for his work on valency. There were also several German assistants with whom he became friendly. Not only his colleague Buff, but also Reinhold Hofmann, private assistant to Williamson, and Hugo Müller, assistant to Warren De la Rue, provided stimuli. Peter Griess, later discoverer of azo-dyes, was another member of the set. The number of Germans declined in the 1860s, but even then the number of Chemical Society members from the metropolitan academic circle continued to increase. In the 1860s there was a boost to the enrolment to the Society from the Royal School of Mines after Frankland succeeded Hofmann. Three or more metropolitan academics were elected annually in seven out of the ten years. In the preceding decade only four years had brought so many of the category to the Society. Though the figures testify to the continuing vitality of academic chemistry in England, the outstanding position of London in world chemistry was transient. Within a few years of Kekulé's visit, Frankland, a vice-president of the Chemical Society, would complain of the relative paucity of British research contributions.[41] Germany had become the undisputed centre of world chemistry.

The Chemical Society had been founded as a coalition under academic leadership; it maintained that character. Nevertheless, the membership data suggest that the Society was not of widespread interest to manufacturing industry, and the manufacturers' presence was only as strong as it was because of the number of industrial chemists with an academic orientation. The balance between their academic and industrial interests cannot of course be known; however, current manufacturing interests by themselves seem to have been a weak affiliating force. On the whole, the academic discipline of chemistry did not prove to be in itself a basis for industrial innovation, even in the chemical industries.

Chemical industry

By the 1840s there was a well established convention of what industries 'Chemistry Applied to the Arts' might be expected to include. An authoritative statement was the work by Friedrich Knapp, Professor of Chemistry Applied to the Arts at Liebig's University of Giessen.[42] It was translated into English and copiously annotated by two well known Giessen PhDs, Thomas Richardson of Newcastle and Edmund Ronalds, the first editor of the new *Quarterly Journal of the Chemical Society*. The first volume of this edition dealt with the products of heating and included coke, charcoal, gas, oils, sulphuric acid, soda and soap. The second volume dealt with the processing of siliceous deposits and included the manufacture of glass, alum and ceramics. Finally the third volume dealt with the manufacture of food, particularly of sugar. Though it did give some order to the proliferating variety of trades that were said to exemplify chemical principles, this academic construction was rather artificial. It associated technologies which had little in common industrially.

This diversity masks the fact that a dominant industry was

emerging in Britain at that time.[43] The manufacture of soda through the Leblanc process developed in the mid-nineteenth century as a chemical industry on an unprecedented scale. With it were intimately connected the manufacture of sulphuric acid (an essential ingredient), bleaching powder (made from waste chlorine), soap and glass (made with soda), caustic soda and so on. The industry grew up where the necessary raw materials – limestone, salt and pyrites – were available. That meant primarily Merseyside, Tyneside and Clydeside. Rather than a few tons per week, the conventional scale of chemical trades, hundreds and even thousands of tons per week were characteristic. A comparably large scale in each of the interlinked stages was necessary to ensure the smooth running of the whole industry. Production of soda ash increased from 72,200 tons in 1852 to 208,000 tons in 1878. In 1861 it was estimated that 10,000 people were involved in the fifty soda works.[44] Similarly sulphuric acid manufacture grew rapidly. Output increased from 380,000 tons in 1865 to 590,000 tons five years later.[45]

The size of the chemical industries was impressive and often cited. Even in the absence of statistics, the scale of plants was often recounted by the Victorians. Their 'scientific' basis was much emphasised but the rhapsody of this prose should not in itself lead us to suppose that the factories were chemical laboratories writ large.[46] Certainly chemistry provided useful skills in analysis.[47] As we have seen industry came to employ chemists to carry out routine tests or alternatively engaged the services of private consultants. However, the role of the chemist in changing, as opposed to monitoring, processes is more problematic.

Patents

A sense of the balance and orientation of innovations in chemically related fields can be obtained from examining the patent literature of the period. Of course not every invention is patented but patents constitute at least a rough equivalent to the journal literature of science. Treated as the tips of the iceberg of invention, patents can be useful indicators of important phenomena.[48]

Between 1830 and 1866 the number of British patents increased spectacularly. The number registered each year increased from less than 200 to over 3,000. This rise was made possible by changes in patent law in the mid-1830s and in 1852.[49] Each of these changes was followed by a jump in registrations. The number registered in 1853 was six times higher than two years before. Aside from such leaps induced by legal changes, the rate of increase in patent registrations was, however, not great. The growth over the fourteen years 1853 to 1866 was equivalent to barely more than 1% per annum. This low rate was recognised in the 1860s and considered by a parliamentary committee in 1862.[50] It was concluded that two factors were at work. First, before 1852 the lack of reference works to existing patents had meant that there was a considerable amount of reinvention that had inflated the earlier figures. Secondly, it was difficult to determine long-run trends because of the fluctuations induced by economic cycles. When times were bad inventors felt less able to bear the considerable cost of taking out patents and keeping them in force.

Though it is difficult to identify peculiarly chemical patents there exists a useful compendium of a category called 'Acids, Alkalies, Oxides and Salts'. This was Class No. 40 of Bennet Woodcroft's *Abridgements of Specifications*, published in the late 1860s with supplements later in the century.[51] The list includes all patents mentioning the names of the chemicals in its title. More than any other single class of patents Class 40 represents a contemporary grouping of chemical innovations. The significance of the class was tested by comparing a comprehensive list of all 1,122 new British patentees cited in it between the years 1830 and 1875, with a list of Chemical Society members compiled from the patentees in all fields in the years 1860 and 1861, and 1870 and 1871. Most of the Chemical Society members who patented in those years would have been discovered from the Class 40 listing. Of the sixty-three who patented in 1860/61, forty-four appeared in the Class 40 list, while thirty-nine of the sixty-two 1870/71 Chemical Society patentees were caught. Therefore examination of Class 40 would seem to yield information about the patenting activities of chemists in general.

The growth of Class 40 followed the overall growth of patents generally. However, the numbers before 1851 were very small, barely exceeding twenty in any year. In 1851 the number was fourteen. Then came the change in law. In 1853 the number was seventy-six. Thereafter, in the years to 1866, systematic growth was negligible.

However, it is not with the notoriously difficult and misleading patent statistics themselves with which we are concerned. It is the patentees who are of interest. Even after restricting the analysis to Class 40 the numbers to be analysed are considerable. Over 1,700 people patented in the class to 1875. (Detailed figures and statistical interpretation are provided in Appendix D) The list of patentees includes both British and foreign, particularly French, residents. The total number of foreign patentees was not trivial; it amounted to 587. Of these more than half, 316, were French. The significance of the French is perhaps surprising, given the accepted wisdom that it was the Germans who were pre-eminent in the chemical industry in the second half of the nineteenth century. However, research now appearing on the dye industry gives a complementary picture of French eminence and suggests that patent statistics can disclose interesting and unexpected phenomena.[52] Parenthetically, it is interesting that the Americans were also very active, perhaps unexpectedly so.

The growth of the patentee community can be seen through an analysis of the 'birth rate' of patentees. The approach has been to examine in detail the dates of first patent (in Class 40) by individual patentees. Communities of British and foreign patentees grew at different rates. For the former their British patent would be the first taken out; on the other hand, since people would normally patent at home first, British patents of foreigners would have followed a domestic claim. After the legal change of 1852 'birth-rates' in the two communities grew according to different patterns. Starting from a lower base, foreign patentees seem to have increased somewhat faster in this area than British patentees – at 7% per annum compared to 5% for local patentees. This discrepancy between the growth rate of British and foreign originating patents, while not discussed directly, coincided with oft-cited concern over greater foreign originality.[53]

We can segment the British community of chemical patentees further. One striking division is apparent from inventors' self-descriptions given on patents. Some patentees styled themselves 'chemists' or variants such as 'consulting' or 'analytical chemists'; one could also perhaps include in the category manufacturing chemists and chemical manufacturers. 'Engineer', a title hardly seen in this category before 1852, thereafter quickly became quite common: an annual growth rate of 12% gives a good approximation. Meanwhile the chemists grew erratically at only 3% per annum, or 4% if one excludes the chemical manufacturers. Though even in 1875 the engineers did not outnumber chemists among first-time patentees in the area, their place was significant and systematically enlarging.

Industrial chemistry

The impression gained from quantitative patent data is complemented by contemporary appreciation. The relevance of chemistry to the chemical industry was continually discussed. The Secretary of the Birmingham and Midland Institute could reiterate this familiar theme even in 1903 in a speech that promoted the institute as a link between science and local industry:

The British Manufacturer as a rule wants something more than a mere Chemist, or if he does take a Chemist from a University he often forms unreasonable expectation, and is often disappointed, the more so because the young fellow is often too optimistic as to his own powers, and so the reaction is the greater. Is it reasonable to expect him to know the working conditions of a large industrial concern? Is he to translate instantly the experiments of his beakers and crucibles into boilers and furnaces?[54]

By contrast to scepticism about the use of the chemist, the engineer was well appreciated. In supporting a proposed association of manufacturing chemists in 1872, 'Black Ash' wrote:

I hope that the Association, if formed, will take in as members those gentlemen of the engineering profession who give a portion of their time to the study of chemical plant. For my part, and as a manager of a chemical works, I think a knowledge of engineering quite as essential as a knowledge of chemistry for all those who superintend chemical works.[55]

The engineer's role in chemical industry had for some time been recognised by the term 'chemical engineer'.[56] By the late 1870s George E. Davis was considering the possibilities of founding a society of chemical engineers.[57] In his *Handbook of Chemical Engineering* Davis argued the increasing importance of engineering expertise in the late nineteenth-century chemical industry. Published in 1905, his *Handbook* has been seen as the founding document of twentieth-century chemical engineering, with its keen appreciation of the idea of the 'unit process'.[58] Davis himself had a career firmly located in the late nineteenth-century Lancashire alkali industry. Born in 1850 and educated at the Royal School of Mines, he took his first job at a Manchester sulphuric acid works before 1870 and in the next ten years worked at several alkali factories. He subsequently worked in the gas industry as a consultant and as an alkali inspector.[59] The famous *Handbook* was based on lectures given to the Manchester Technical School in 1887. Davis argued that physics and mechanics were as much part of the theoretical basis of large-scale chemical operations as chemistry. He had noticed that the specifications of chemical patents since 1850 had become much more complete in the description of equipment. If one had to make a choice between a knowledge of engineering and of chemistry, the former would have to take precedence. As a result chemistry by itself, even when honoured with the title 'applied', was inadequate for the understanding of industrial processes. On the other hand there were general principles linking different branches of manufacture. He argued that:

Chemical Engineering must not be confounded with either Applied Chemistry or with Chemical Technology, as the three studies are distinct. Chemical Engineering runs through the whole range of manufacturing chemistry, while Applied Chemistry simply touches the fringes of it and does not deal with the engineering difficulties even in the slightest degree, while Chemical Technology results from the fusion of the studies of Applied Chemistry and Chemical Engineering, and becomes specialised as the history and details of certain manufactured products.[60]

Though Davis's codification was new, much of his wisdom was distilled from existing practices. As a general rule, in the mid-nineteenth century a diversity of separate chemical technologies each had their own language, concepts and problems. At the same time, whether one looks at the work of the engineers or of the chemists, the concentration of chemical patents on a few subjects is unmistakable: soda manufacture and by-products from it, including caustic soda, bicarbonate of soda and bleaching powder, brine evaporation, fertiliser manufacture including ammoniacal salts and superphosphate, and the profitable utilisation of gasworks waste. Each of these branches of manufacture characteristically involved reactions that shared a very complicated chemistry. Hardie has pointed out that it was only in the 1930s, after more than a century of manufacture, that the chemical nature of bleaching powder was elucidated, and even then the chemical kinetics of the Leblanc black ash reaction remained unclear.[61] Therefore engineering, rather than chemical, ingenuity was often called upon even in the soda industry.[62]

In 1861 three eminent Manchester chemists, including Roscoe, reported to the British Association on the improvements in soda manufacture over the previous decade.[63] During that period soda manufacture in south Lancashire and Cheshire had tripled. Despite constant efforts, no fundamental changes in the principles of the

Leblanc process had been effected. On the other hand there had been four major engineering changes. Waste heat from one side of the process was recycled to facilitate evaporation of liquids elsewhere. A series of tanks enabled much improved lixiviation of black ash. The introduction of the revolving furnace meant that the hardest manual work of stirring was mechanised. Soda ash was packed into casks by machinery.

This is not to say there were no significant chemically based innovations. However, their implementation required careful engineering. The 1861 report had pointed to the recent important discovery that black ash solution contained up to 30% sodium hydroxide. Other chemists had confirmed the analysis. From this the dye manufacturers Roberts Dale & Co. had found a new elegant way to manufacture caustic soda. By slowly evaporating the liquors, other constitutents could be persuaded to crystallise out, leaving the caustic soda solution. If the evaporation were carried out in closed vessels one also obtained superheated steam for motive power or for heating purposes. Boiling over of the strong lye could be prevented by an ingenious boiler design perfected at Gaskell & Deacon's works.

Since the soda industry was threatened by chronic overcapacity, diversification into other products such as caustic soda was crucial. But in practice the technical problems almost outweighed the commercial advantages. A manufacturer wrote in 1876:

In truth caustic making is a matter we have to be hopeful about & bear it as we can. It takes a heap of steam, which we cannot measure & is a tangle of unknown quantities. We are just now wanting to find a way of dealing with the caustic sludge, & the caustic fished salts – both very awkward customers ... I am sure people who make caustic only must do better than we do.[64]

This moan came in reply to a query from the Oldbury soda manufacturer A. M. Chance about the market for caustic. A carefully drawn up proposal for investment in a fifty ton per week plant followed nevertheless.[65] The potential profitability was emphasised, based on the growing market for soda, particularly for coal tar processing. Chance investigated the price levels over the previous ten years and calculated that current costs of production were well below even the depressed price then prevailing. Although the proposal did not refer to the current state of other products of the firm, the distress of the soda manufacturers at the time can be inferred from an 1876 letter by Henry Brunner, who worked at one of the largest soda works.

I met several of the Newcastle men in London. They as well as the Lancashire men agree that they are losing money & don't know where it will end.[66]

In fact Chance's careful calculations were misleading. The previous ten years provided no guide to future prices, and within another decade prices had halved again.[67] So alternative products seemed once more attractive.

The portrait of an intimate amalgam of engineering and chemistry drawn by the Manchester chemists for the British Association was reflected in the patent literature. The process of evaporating brine to give salt, scientifically trivial as it was, consumed a lot of energy. Many patents, particularly by engineers, dealt with the problems of economising, for instance by using superheated steam. Analagous problems could be found in sugar refining, and several patents refer to the evaporation of saline and saccharine solutions despite the disparity of industries involved. When, in 1874, the newly established Society for Scientific Industry held its first exhibition in Manchester, the subject was 'appliances for the economical consumption of fuel'.[68]

The Society had grown out of the attempt to found an association of chemical manufacturers in Lancashire (supported as we have seen by 'Black Ash'). The founding spirit was Frank Spence, son of the pioneer alum manufacturer, Peter Spence. Spence based his call for a new society on the need for better communication within the chemical industries. As an instance of the need he cited his wish to know whether the Coffey still, invented for the distillation of alcohol, had ever been applied to ammoniacal liquor distillation.[69] Alum manufacture, in which he was involved, in fact illustrated well the engineering dimension of the chemical industry.

Alum used as a mordant for dyes and for flocculating sewage was traditionally made by calcining heaps of shale containing pyrites and aluminium salts, soaking the result in tanks to extract the aluminium salts and crystallising with potassium or ammonium salts (originally obtained from urine). A greatly improved process was developed by Peter Spence, who had begun his career as chemist to the Dundee gas works. He obtained the aluminium salts from the shale by extracting with hot sulphuric acid and by adding ammonia from gas works liquor, he crystallised ammonia alum. In the early 1860s Peter Spence purchased rights to all the ammoniacal liquor from Manchester gas works.[70] Shortly after, he set up similar arrangements in Birmingham and Hull. Millions of gallons of foul liquid were involved, millions of gallons that the works were only too willing to get rid of. The effort to develop new methods of processing was considerable. It was led by

Frank Spence, who had studied mineralogy at Owens College, studying chemistry with Edward Frankland. His assistant was J. C. Bell, the alumnus of the Royal School of Mines, where he had studied under Hofmann. Despite this array of chemical expertise, as Spence's query about the Coffey still indicated, the problems were largely those of apparatus design. Spence had to find a way of continuously distilling large quantities of recalcitrant ammoniacal liquor. By contrast to the engineering problems he faced, Spence's notebooks suggest that the chemistry was fairly simple.

Partly as a result of Spence's work ammonia became a major product. In a reference list of significant patents assembled by Chance in 1885, twenty-eight patents dealt with producing ammonia, the subject with by far the largest number.[71] Production of ammonia for ammonium sulphate was 40,000 tons by 1879 and 117,000 in 1889.[72] The utilisation of ammonia led the way in the use of gas works by-products. Tar, like ammonia, had been used since 1830 but processing for rubber solvents and creosote had been carried out only on a small scale. Then in the '60s and '70s tar production was expanded mightily. In 1870 it was 117,000 tons and in 1880 640,000 tons. Often the tar was distilled by independent works but the gas industry took an increasing interest. In 1879 the Gas Light & Coke Company opened the by-products plant at its giant Beckton works.[73]

When we think of innovation in the chemical industry at this time, the dye industry is often referred to. There was some interest, clearly. After Perkin's discovery of mauve in 1856 there was a spurt in the early '60s but excitement died after the initial reaction had been exploited, until the patenting of synthetic alizarin in 1869.[74] The dye industry was certainly very chemically sophisticated. If academic chemistry could have relevance, it was here. The coal tar dye industry employed several fine chemists who tackled complex chemical problems. Royal College of Chemistry students came to dominate the field in Britain for a time. The notebooks of J. G. Dale show that careful experiments were carried out.[75] At the same time there was still considerable attention paid to the modification of natural dyes. In the report to the British Association of 1861 far more attention was given to the modification of natural substances such as madder and guano than to the coal tar dyes. On the other hand, dye manufacture was relatively small compared to soda. Even Roscoe, who saw the synthesis of alizarine as a perfect example of the application of chemistry, recognised that compared to the £2 million soda industry, dyeing and calico printing were relatively small industries worth only an annual four to five hundred thousand pounds.[76] Not until the late nineteenth century did even the massively successful German dyestuffs industry exceed the magnitude of the British soda industry.[77]

The engineering difficulties of the chemical industry meant that only a small proportion of the innovations were made by self-styled chemists. This was reflected in the patenting achievements of both Chemical Society members and alumni of the Royal College of Chemistry. Of the former, only about 10% patented in Class 40 and a similar proportion pertains to the latter. In other words, in industry, science and patentable innovation were only weakly connected. Roscoe recognised this when he predicted that his pupils, largely destined for the alkali industry of Lancashire, would work as analysts and quality controllers but made no mention of innovation. Although most of the students of the Royal College of Chemistry also went into industry they patented little. Apparently they too were employed for the routine work envisaged by Roscoe.[78]

Specialisation

Nor were industrially employed chemists major contributors to the disciplinary literature of chemistry.[79] The main industrial chemical activity, analysis, was rhetorically identified with research by the academics. In practice, however, the orientations of the men involved in the two activities were quite separate. A new journal entitled *The Analyst* began publication in 1876 expressly for the practising chemists. It argued that the first volume of Roscoe and Schorlemmer's *Treatise on Chemistry* proved that its authors were incompetent and as for the *Journal of the Chemical Society*, it contained little of interest:

The next sixty pages are chiefly devoted to abstracts on organic chemistry, which we pass over with the remark that a very undue proportion of the space is devoted to papers, the authors of which claim an intimate acquaintance with the exact position of every atom in even the most complicated compound. This may be very interesting to chemists who, like the authors of these papers, fancy themselvs hail-fellow-well-met with every atom under the sun, but it is somewhat tedious to ordinary chemists not on the visiting list of either atoms or molecules.[80]

The tensions between practising chemists and their academic colleagues expressed in *The Analyst* were part of a more general specialisation in chemistry. From the mid-nineteenth century the academic science was becoming formally divided into different specialties.[81] The subjects of 'inorganic', 'organic' or 'physical' chemistry came to define the range of even the most talented individual's expertise. Research problems within the specialties were very different. Organic chemistry was at this time often considered the most exciting as the developing understanding of molecular structure and valency theory made possible new syntheses and structural determinations. In Britain the first Chair to be restricted to a single chemical specialty was the professorship of organic chemistry created at Manchester in 1874.[82] In inorganic chemistry meanwhile the periodic table and the developing techniques of spectroscopy defined leading research fronts. Physical chemistry as a specialty did not flower until the 1880s though even in the 1860s rates of reaction were being investigated by men such as Harcourt and Esson at Oxford and by Guldberg and Waage in Norway.[83] Thus divides were appearing not only between academic and industrial skills but even within the corpus of 'pure' chemistry. At the same time there was a rhetorical emphasis on the integrity of the discipline and its common centre.

Chapter Five:
Enquiries

By the late 1860s the coalition between academic, manufacturing and consulting chemists established two decades earlier was losing stability. It had sustained the first stage of institution building, but the very class of academics thereby created was now shaping the discipline itself. The influence of the new pressure group was all the greater because the structure of education as a whole was in ferment during the 1860s. Expressions of mounting interest in technical education encouraged government moves in which scientific entrepreneurs found welcome opportunities to increase support for their own institutions. The scientific professoriate portrayed themselves as disinterested spokesmen for the country's future in tones that still resonate a century later. Perhaps this modern appeal has deterred detailed investigation of their thought. However, the analyses of the 1860s deserve to be understood as distinguished achievements in the integration of personal calculation on the one hand with visions of the long-term national future on the other. They were formulated in the course of a series of portentous debates about government science and education policy. Large sums of money and the national future seemed to be at stake. Appealing to the government's interests, scientists made explicit the concepts of pure and applied science which had developed in the previous quarter-century. In the political manoeuvres and public enquiries of the late 1860s and early '70s, the distinction between the two levels of science acquired an enduring institutional and political reality. Moreover the meaning of the categories was indirectly debated through prolonged discussions over the better institutional form for the teaching of science: polytechnic or science school.

This chapter concentrates on the political components of the enquiries, on the testimony before them and on the final reports. Chemists whose work and opinions have been encountered in earlier chapters – Playfair, Frankland, Roscoe and Williamson among them – were leading witnesses. Through their testimony and the committees' reports one can see the transition from local experience to institutionalised assumptions.

Behind the educational initiatives of the 1860s and '70s was an increasingly well established faith in science both as useful knowledge and as liberal education. After the expansion of metropolitan science teaching in the 1840s, both Oxford and Cambridge brought science subjects into their schemes of 'liberal education'. The Honours School in Natural Sciences at Oxford was established in 1850 and the Natural Sciences Tripos, initially as a non-degree course, at Cambridge a year later.[1] Following these universities' initiatives the 1860s saw a crisis of confidence about the structure of liberal education at the public schools. In the age of Darwin, the ideologists of science, most notably Huxley, Spencer and Mill, argued that a liberal education would have to be science-oriented. The Clarendon Commission on the Public Schools in 1862 was impressed by the testimony of men such as Lyell and Faraday. They reiterated the theme that science would provide appropriate broadening of the mind, being as 'liberal' as the classics. Science would at the same time enable the upper classes to understand the principles of industry which was enriching the middle classes.[2] Five years later, Farrar brought out the famous *Essays on Liberal Education* which highlighted the significance of science.[3]

Historians have emphasised that reform at the great public schools was marginal at the time.[4] However, debates conducted during the 1850s and '60s did reorient the philosophy of liberal education. For the middle classes seeking such an education for their sons, science would be an appropriate subject for study, useful for the development of character and as preparation for later life. From 1861 it was possible to acquire a degree in natural sciences at Cambridge. The educational value of science was further endorsed by the examinations of the reformed civil service. Middle-class education was the subject of a major public investigation by the Schools Inquiry Commission, which reported in 1868. The place of science in the secondary school curriculum was an important issue for the commission. It conducted detailed surveys of what was included in European curricula in general, and in British science curricula in particular.

As for utility, the attraction of science as useful education was highlighted by growing fears that Britain was 'declining'. As early as the Great Exhibition of 1851 Playfair had cited shortage of technical education as the cause of an impending British decline. Few disputed the virtues of the call for more 'technical education' for the working classes. It could be appropriately funded out of the profits of the Great Exhibition. However, the use of public finance would create a new and contentious area of public responsibility. Moreover there were conflicting views as to what technical education should include.

Definitions of technical education

Right at the beginning of the 1850s it became clear that publicly supported technical education could not include training in vocational skills. In the first thrust of the technical education movement that followed the Great Exhibition, Lyon Playfair with the connivance of Henry Cole, Prince Albert and the government all endeavoured to set up the new system of central college and provincial science schools described in Chapter Three. Despite pious demands for technical education from regional Chambers of Commerce in the early 1850s, there was little industrial support.[5]

When the Society of Arts solicited interest in a new scheme of technical education in 1853, the few manufacturers who did answer showed an ambivalent attitude to the industrial utility of education.[6] Of the ten leading industrial correspondents, three were calico printers. Two out of the three, Walter Crum and John Mercer, were themselves distinguished founder members of the Chemical Society, the third, Richard Fort employed several chemists. The three paid homage to the importance of the intelligent working man. Crum, however, felt that while 'pure' science could be taught in schools, applied science had to be kept to the workshop. Mercer disclaimed an understanding of 'abstract science' and urged that pupils be taught only subjects relevant to local industries. He favoured the branches of chemistry relevant to those manufactures. Richard Fort took a similar view. Their chemistry would steer the pupil between the Scylla of useless knowledge and the Charybdis of unteachable technique.

The opposition by manufacturers to the teaching of trades was characteristic and repeated in every discussion of the subject. This was emphasised even by supporters of 'technical education'. In an 1868 article entitled 'What is True Technical Education?' *The Economist* explained:

Technical Education must mean, then, something very different from apprenticeship in Government Polytechnic schools. Each one of our ten thousand trades must still remain an unwritten mystery, and must pass by tradition from one workman to his apprentice; but there is still very much for us to do. It is in the preparatory stage of education that we can raise the intelligence and knowledge, and, therefore, the ultimate skill of the artisan ... It is to *efficient school education, plus apprenticeship,* that we must look for a supply of highly-skilled and intelligent artisans and foremen ... Science is now the recognised basis of improvement in industry.[7]

At face value this ode to the apprenticeship system would seem to have been misplaced. The time was full of complaints that apprenticeship in Britain was in danger of disappearing, that apprentice training in Britain was inferior to that on the Continent, and that it was in any case outdated. The workmen who reported on the 1867 Paris exhibition were impressed by the system of technical education they saw in France.[8] But the emphatic defence of the mysteries of trade knowledge and craft skill had its commercial as well as craft connotations. Donnelly, the effective head of science in the Science and Art Department, was to complain that manufacturers were loth to allow their workers to gather for trade education in case they exchanged trade secrets.[9] Such anxieties had been long prevalent though their end was seemingly always in sight. Even in the 1840s *The Chemist* had forecast the end of obsessive secrecy as a bar to education.[10] Another objection of manufacturers to formal education was explained in a minute to the President of the Council and echoed by Henry Cole, head of the Science and Art Department. It was suggested that 'Real technical education–the teaching of a trade or art itself on scientific principles – necessarily entails workshop practice'. However, such workshops would produce goods that would compete unfairly against private enterprise. 'A state pottery school and Messrs Minton could not exist side by side.'[11]

The opposition of manufacturers to trade education was a fundamental restriction on the definition of 'technical education'. Prescriptions for technical education therefore fell back on science which combined knowledge of principles with personal improvement.[12] Science, it was argued, enabled the worker to understand the theory of his labours and to think better. Thus when the Science and Art Department instituted its examination scheme in 1859 for subjects 'In a greater or lesser degree to have a bearing on industrial occupations', sciences, particularly chemistry, had a primary place.[13] It was not just that this was what government wanted to offer. There was a considerable demand. In 1856 a memorial signed by 5,000 'working men belonging to the Royal Polytechnic, the

London Mechanics' and Other Institutions in London' had appealed to the Vice-President of the Council for government funds. These were to be used to support the study of the principles of science.[14]

The discussions following the Great Exhibition established that 'technical education' was principally scientific education. They also firmly established a separation of technical and elementary education. This was reflected in administration. The Science and Art Department was initially within the Board of Trade and even when transferred to the Committee of Council for Education was organised separately from elementary education. The Revised Code for elementary education established in 1861 rewarded teachers for such a restricted curriculum that science was excluded. Scientific education would be reserved for adults and separated from the teaching of the fundamental 'three Rs'. Reading, writing and arithmetic became the only grant-rewarded subjects under the Revised Code. The income, status and responsibilities of teachers fell as their range of subjects was constricted. The numbers entering the profession quickly fell as the number of pupils was rising.[15] Not surprisingly the training of the teaching profession was a recurrent issue in the 1860s. And in spite of the formal division between the two State-supported sectors, elementary and technical education, in practice many of the elementary school teachers by day enhanced their income by teaching Science and Art classes by night.[16] Inevitably the question of teacher training came to involve scientific education.

The polytechnic and the science college

By the mid-1860s, then, there was political concern about the three areas of education in which the State was involved: elementary education, technical education and teacher training. In the context of the re-evaluation of political issues after the second Reform Act attempts were made to bring them into one system. This came to entail consideration of an unprecedented State involvement in secondary and higher education. Two outcomes were particularly relevant to the relationship between science and practice. Firstly a science curriculum which clearly distinguished between preparatory pure and professional applied science was identified as peculiarly worthy of support. Secondly, aspiring teachers for whom even professionally oriented training was actually the study of the discipline itself came to be seen as the prime constituency to be served. These conclusions, reached by the mid-'70s, were the result of prolonged political and institutional negotiation. During the early 1860s they would perhaps have seemed surprising. For in the aftermath of the successful implementation of the payment-by-results system by the Science and Art Department, the most popular idea for the next step had been a technological university or polytechnic on French lines. Particularly esteemed as a model was the Ecole Centrale des Arts et Manufactures in Paris which was specially intended for aspiring manufacturers. This was animated by the concept of 'la science industrielle', expressing unity in a curriculum that stretched from abstract theory to particular applications.[17]

There was already the School of Mines, which though rather diffuse could be portrayed as an English analogue to Continental mining schools. In 1861 a Treasury committee decided that this should focus even more on mining. The School should discontinue courses in various peripheral subjects for occasional students (with the exception of chemistry). It was suggested that diplomas be awarded. Regrettably University of London degrees could not be made available, since there was a classics requirement for matriculation. Therefore the Treasury committee suggested that an autonomous body be set up to give qualifications to graduates, and that its status be reflected in the title, which should be the Royal School of Mines.[18]

A second and even more specifically French-influenced institution was the Royal School of Naval Architecture, which opened in South Kensington in 1864.[19] Though it had initially only a temporary building, a spectacularly magnificent permanent home was under construction through the 1860s. It was the financial responsibility of the Admiralty but was administered by the Science and Art Department, which would take financial responsibility for any non-naval pupils who attended and had supplied the building. It occupied premises belonging to the department. The School arose in the first instance out of a scheme proposed by the engineer John Scott Russell to the Institution of Naval Architects in 1863. The naval scare of the early 1860s plus the debates over the transition to iron and steam tended to bring support to Russell's plan. Russell was an intimate and former collaborator of Henry Cole and definitely discussed the plan with him in 1863.[20] It was Cole who took the plan further, getting Admiralty support and setting up the mechanism for launching the school. Its landlocked position is then hardly surprising. Having failed to win over the geological interest in the

1850s once Murchison had been appointed as Director of the School of Mines, Cole attempted to relaunch his idea of a central institution at South Kensington. His promotion of this unlikely spot for a naval school may be seen as the first step.

Russell was unpopular with the First Lord of the Admiralty, whose naval policies he opposed. So what was seen to be his scheme could not be used outright. Instead, the Committee of Council on Education, presumably at Cole's behest, appointed Captain Donnelly and Joseph Woolley to frame a scheme in consultation with the Chief Constructor of the Navy. Woolley had been head of a now defunct naval school in Portsmouth and the Chief Constructor had been a pupil there. As part of the preparations, they were instructed to visit the Ecole Impériale du Génie Maritime in Paris. This became their model, as indeed it had been the model of Russell, who had sent his son there. The course they proposed was very much a polytechnic course along the lines that Playfair would have liked to have been able to develop for chemistry. The course began with a thorough grounding in pure and applied mathematics, and then proceeded to the theory of practice (including some practical work) of various aspects of shipbuilding and marine engine design. It also included study of the principles, with laboratory work, of the basic sciences of physics, chemistry and metallurgy. Very importantly, students were meant to spend only six months of each of their three years on course-work at South Kensington; the remainder of the time was to be spent actually at work in a shipyard. It was envisaged that the bulk of the pupils would be sent by the Admiralty, although private shipbuilders were also expected to send students. It was even suggested that during the practical work phase of each year Admiralty and private students might swop venues to enhance experience and communication. Occasional students were also to be welcomed.[21]

Gradually emerging in contrast to the polytechnic ideal was the science college. At first this was conceived as a training centre that would prepare men not for only a single industry, but more efficiently for them all. The idea was first explored in Ireland. In the 1840s a tripartite scientific institution had been established in Dublin on a pattern similar to the Geological Survey and its school and museum in London. Dublin had a Museum of Economic Geology with its own classes and a link to the national Geological Survey. In London, as we have seen, Playfair tried and failed to break the geological hegemony. A similar more successful attempt was made at the same time in

Ireland.[22] Sir Robert Kane, an eminent chemist, became head of a retitled 'Museum of Irish Industry' with some teaching responsibilities. At the same time Kane was Professor of Chemistry at the older Royal Dublin Society, which also offered classes with government funding. When in the early 1860s the Royal Dublin Society bid for a much increased level of government support, Donnelly manoeuvred the Treasury into insisting on an enquiry into the whole situation.[23] The first group to report suggested that the Royal Dublin Society take responsibility for the existing teaching functions of the Museum. After considerable local pressure, this was rejected by Parliament and a Select Committee reconsidered the matter.[24] Finally the Committee of Council on Education decided that while a government-backed system of specialised professional colleges such as was emerging in England would be desirable in principle, it was not practical. The demand for employment hardly sustained the School of Mines and would certainly be inadequate in Ireland. Instead, in September 1865 it was proposed that the Museum of Irish Industry should become:

A College for affording a complete and thorough course of instruction in those branches of Science which are more immediately connected with and applied to all descriptions of industry including Agriculture, Mining and Manufactures and that it should in this way supplement the elementary scientific instruction already provided for by the Science Schools of the Department towards the training of teachers for which schools it may also give considerable assistance.[25]

The curriculum for this school was devised by a commission of eminent scientists, including several from the School of Mines in London, notably excluding geologists. Chaired by the eminent amateur Irish astronomer the Earl of Rosse, its eighteen members included the chemists Robert Kane, Edward Frankland, A. W. Hofmann and Lyon Playfair. The polemicists for science T. H. Huxley and John Tyndall were also members. The scientists confirmed the politicians' judgement. They argued that the content of the curriculum should not in itself be practical. Rather it should 'impart a sound and thorough knowledge of those branches of science which may be so applied ... but practical subjects when capable of being rendered illustrative of scientific principles should in all cases be introduced in the course of instruction'. A three-year curriculum was suggested, including a common first two years extending over all the sciences with a final year dedicated to an applied specialty. Students still had the option of being occasional students. The proposed system explicitly differentiated between the 'pure' science of the first two years and the 'applied' science of the final year. There was even to be a separate Chair of applied chemistry. Hofmann's two-part course at the Royal College of Chemistry had therefore now been translated into 'pure' and 'applied' components.[26] Though there was ritual reference to the polytechnic, the school was far less closely connected to practice than was the School of Naval Architecture.

The Dublin College and the School of Naval Architecture each represented new forms of organisation for science education in Britain. Their example and the more tangible, albeit slowly rising, building in South Kensington constituted important intellectual and physical resources as the national interest in technical education reached a new peak in 1867 and '68.

The Samuelson Committee: origins

In May 1867 Playfair, reporting from Paris, wrote a scathing letter on the second-rate performance of British exhibitors at the Paris exhibition. Addressed to Lord Taunton, the chairman of the Schools Enquiry Commission, the letter marked the beginning of a revival of what Disraeli called the 'Albertine' movement for technical education. Following Playfair's blast, Taunton investigated the views of educationalists and pundits about technical education and evaluated consular reports on Continental methods.[27] At the same time, the reforming MP and iron manufacturer Bernhard Samuelson sent a report to the government on his detailed survey of technical education overseas and gave his own proposals. He had seen very successful apprentice schools on the Continent, but felt them not appropriate to British conditions. Instead, expressing an argument to which he was to return frequently, Samuelson suggested that enhanced scientific education would be the answer. Better teachers were an essential prerequisite and training them should be the responsibility of the government. The best pupils in elementary education should be sent through the Science and Art classes to the School of Mines in Jermyn Street or to local colleges to receive advanced education to fit them to be teachers. At the same time, these local colleges could also train future manufacturers. They should be built up with local effort supported by government assistance. Thus

Samuelson argued for a unified State system offering through science a combination of liberal and of practical education that would satisfy both the working and the middle classes.[28]

Nor were Parliament and government slow to see the advantages of integrating the two hitherto parallel movements of scientific education for the working and middle classes. Following an unsuccessful opposition attempt in early summer 1867 to promote an elementary education Bill, the Conservatives put forward their own measures.[29] In October, when Disraeli was in Edinburgh to receive an honorary degree, he announced that improvement of education was on his programme. To the Tory faithful assembled in the Corn Exchange he lauded the existing system of technical education. The next evening, speaking to working men, his emphasis was different. He proclaimed that the government was intent on helping the British worker stave off foreign competition by contributing to technical education.[30] That winter the government planned changes to both technical and elementary education that, if not revolutionary, would be a major step in the evolution of the education system.

On 12 November 1867, four days before Samuelson's officially separate missive on technical education, Donnelly, head of the Science section of the Science and Art Department under Sir Henry Cole, had written a minute on a new plan for 'Scientific Instruction'.[31] Within two weeks Cole submitted a minute integrating Donnelly's proposal into a comprehensive scheme. It united 'primary or elementary education' with 'secondary or technical instruction' and with higher education, including scientific instruction for future teachers and manufacturers. The scheme would assist local enterprise but would entail the support of considerable resources from both local and central authorities. A first step towards establishment of the scheme was taken just before Christmas that year, by provision of scholarships for pupils in elementary schools to local science and art schools and further scholarships for successful pupils to colleges of science. The official 'minute' did not specify which colleges, though the explanatory text referred to the School of Mines and the Royal College of Science in Dublin.[32] It was intended that there should be at least four such science colleges linked to local industry in the provinces. This ambitious plan was abandoned by the government. Cole had proposed that elementary education should be funded from local rates, a proposal that foundered on sectarian interests. Instead, Disraeli proposed a less controversial scheme for elementary education. This was put to the House of Lords in March 1868 but it ran out of time and was not pursued. The technical education side did not even reach Parliament, allegedly for lack of money. However it is also true that Lord Montagu, its sponsor in the government, was not considered sound.[33]

In the absence of further government action the plan was kept on the political agenda by a variety of parallel extra-governmental enquiries and proposals. As the Minister responsible for the Science and Art Department within the government, Lord Montagu formally asked Chambers of Commerce for their opinions as to the damage being experienced owing to lack of technical education. They obligingly replied in tones that ranged from mild to apocalyptic. James Hole of the Yorkshire Union of Mechanics' Institutes and chairman of the Chambers of Commerce, wrote to Montagu proposing a system that looked very similar to the one under consideration by the government.[34] At the same time, in January 1868, the Society of Arts organised a large meeting to consider technical education, a few days after another conference on elementary education. The meeting decided to appoint a committee on technical instruction to advise on actual curricula.[35]

It is not surprising that Playfair in Edinburgh and Roscoe in Manchester were, that January, preparing expensive claims for support from the government. Samuelson gave tactical advice to the Owens Extension Committee on the best way to phrase an appeal for funds. He explained to the Manchester Extension Committee the points that would need to be stressed in a submission to the Treasury of their request for £100,000.

1. If shown that the community would be prepared to contribute to the support of a college of science.
2. That such an institution would be available not for Manchester alone but for a widely extended manufacturing community.
3. That in return for Government aid, gratuitous, or partially gratuitous, instruction might be given to persons in training as teachers.[36]

Clearly Samuelson felt that the views he had expressed in his proposal of November might be realised through Owens College. On 25 January 1868 *The Economist* had argued for what was effectively Montagu's scheme: a system of primary education, science schools and colleges and teacher training in science to complement apprenticeship.[37] Two weeks later the calico printer Alfred Neild on behalf of the Owens College appeal published a letter in *The Economist* entitled 'Technical Education'.[38] He argued that attention should be given to the education of the *'chiefs* of industry'. Their training should be in the new science colleges such as the one Owens was trying to expand into.

On 24 March 1868, the same day as the government's elementary education Bill was being considered in the Lords, Bernhard Samuelson, actually in the Opposition but with Montagu's support, proposed an investigatory sub-committee to the Commons. Samuelson suggested that a Select Committee investigate the instruction in theoretical and applied science to the industrial classes.[39] This formulation was explained at the beginning of his speech. The phrase 'technical education' had been avoided because it had two meanings – education in the factory and education in the school. Samuelson was concerned only with the latter. He argued that a polytechnic system of technical education was called for. He was answered by Montagu in a long, well argued speech. Montagu replied that indeed technical education was now being demanded to help fend off foreign competition. This alone would be a significant justification. But its primary function would be to improve the character of the nation, rather than merely its wealth. Moreover education would give opportunities to workers who in that year of the founding of the TUC would seek other ways of advancement. 'Hitherto the means sought by working men to rise in life were unions to obtain less work and more pay. He believed that they now saw a better way; they were aware that by knowledge they would command a superior position.' He reminded Parliament that the idea of special training had failed in the early 1850s. 'We should therefore teach the sciences which are applied to trades, but not the trades to which sciences are applied.'[40] His system would enable intelligent boys to rise as far as their merit would take them, and would encourage national cohesion. As a result of the debate, Parliament set up a Select Committee. This 'Samuelson Committee', as it was known, after its Chairman, Bernhard Samuelson, who had initiated the parliamentary proceedings, began work very quickly. It was ordered on 24 March and heard its first witness on 23 April. Proceedings thus overlapped those of the Society of Arts Technical Instruction Committee that had been set up in January (both committees reported in July), and it was already sitting when the Schools Enquiry Commission reported.

Samuelson Committee: testimony

The Samuelson Committee made an important contribution to contemporary discussions on educational reform.[41] Testimony was taken from civil servants, scientists, teachers and industrialists. The committee provided a forum for the policy-makers within government to explain their views, and an opportunity for chemists and other scientists to show the relevance of their philosophies to the government's policies. The witnesses' testimony provide a glimpse into the emerging relationship between government and academics. It is clear that there was considerable private consultation in advance of testimony. Cole recorded in his diary an unsuccessful attempt by Samuelson to get him to argue the importance of a separate science department independent of art.[42]

Though the committee had been set up under the shadow of industry's needs, the training of science teachers came to be the focus of discussion. Altogether fifty-eight witnesses testified. The principal participants were Henry Cole, Secretary of the Science and Art Department, and Captain Donnelly, the Official Inspector for Science. Three chemists, Lyon Playfair, Henry Roscoe and Edward Frankland, were leading academic witnesses. The administrators stressed the importance of training science teachers; the professors stressed the needs of their colleges.

Cole and Donnelly were the first witnesses called. Having described the workings of his department's examinations, Cole pointed to what he saw as the major problem in the provision of scientific instruction for artisans: finding and retaining good teachers. Well trained teachers were attracted to industrial jobs as soon as their reputation became known in an area. Though an evening teacher could earn something of the order of £150 p. a., the average income was more like £50. By contrast Frankland and Playfair both spoke of £300-£400 p. a. for works chemists.[43] It was not just that industry offered better salaries, but also that teaching to artisans under the department had to be done in the evening. Constant reteaching of elementary material was often mere drudgery.

On the question of State aid for education Cole was equally clear, and he reiterated the views he had expressed to the Committee of Council on Education the previous December. He wanted to see the voluntary system of education retained, with State support limited to the poor. (The Science and Art Department only awarded payment by results for successful artisan examinees, defined as those whose annual income was less than £100 and who therefore were not subject to income tax.) The middle and upper classes, Cole argued, should be eligible to attend classes receiving State support but they should pay economic fees for the privilege. Furthermore, emphasising the voluntary principle, Cole laid stress on the importance of local initiative; it was up to a locality to put forward proposals for scientific education and then to make an approach for central support. Emphatically, it was not the duty of the State to take initiatives in particular areas, or to direct education.[44] Lastly, Cole reiterated the definition of technical education as the 'teaching of those general principles of science which tend to improve the industry of a country'.[45] Here he was prompted by his sympathetic questioner, his political master, Lord Robert Montagu.

Cole's second-in-command, Captain Donnelly, was full of praise for the system of State aid run by the Science and Art Department because of the range of institutions it supported: night classes in mechanics' institutes, day classes in mechanics' institutes, night classes in elementary schools, secondary science schools with both day and evening classes, classes taught under the Free Libraries Act, and classes taught under the auspices of private firms.[46] On the other hand, he thought it unreasonable for the government to give special training to science teachers since it would be difficult to find them sufficient remuneration. Donnelly's ideal science teacher was one who did something else in the day and turned to Science and Art Department teaching in the evenings. In particular, he thought that students training for elementary teaching should receive additional instruction in science so that they could do supplementary evening work as well.[47] Questioned by Montagu, Donnelly declared himself in favour of central colleges for science instruction, on the model of the Dublin Royal College of Science, to be located in London and (believing even more than Cole in local initiatives) perhaps also in Edinburgh or Manchester (the homes of Playfair and Roscoe respectively). Their purpose would be to train teachers as well as to impart general scientific instruction.[48]

In their testimonies the academics Playfair, Roscoe and Frankland each laid their own emphasis but they were all anxious to express their support for a new integrated education system. Playfair admitted, as Frankland did later, that there was no proven need for an expansion of tertiary training. Supply and demand seemed to be more or less in balance, as 'it is only the more intelligent and advanced manufacturers who make the demand' (for trained technical people).[49] Nevertheless he gave his own view of applied science. Excepting Oxford and Cambridge, which he thought ought to devote themselves to 'pure science rather than applied science', he recommended that the other tertiary institutions be developed into polytechnics. As for the location of the polytechnics, Playfair favoured the idea of improving or reshaping existing institutions rather than creating new ones; hardly surprising, in view of his current ambitions for Edinburgh.[50]

Playfair was not questioned on the matter of training teachers, but that was an issue which dominated Frankland's testimony. He argued for a special course of training for science teachers in connection with the Royal School of Mines.[51] Roscoe too came out strongly in favour of training teachers as a principal role for a tertiary science college. 'I think at the present time that [the training of science teachers] is the great work which we, in the science department of the College, have to do; and I think that that alone would be a return for the endowment of such colleges [by the State].'[52] He noted particularly a great lack of science teachers in the Manchester area. Again, since we know of Roscoe's previous coaching from Samuelson, it is clear that Roscoe was aware of the political resonance of his opinion.

Samuelson Committee: report

The committee's report noted that the witnesses were convinced that:

a knowledge of the principles of science on the part of those who occupy the higher industrial ranks, and the possession of elementary instruction by those who hold subordinate positions, would tend to promote industrial progress by stimulating improvement, preventing costly and unphilosophical attempts at impossible inventions, diminishing waste, and obviating in great measure ignorant opposition to salutary changes.[53]

There had been less unanimity about the extent to which the State ought to provide such education. Some witnesses had argued that high premium payments for apprenticeships were evidence that private enterprise would be sufficient, while others had argued that education developments were only ever taken advantage of after the State had taken the lead. Though the witnesses had agreed that the State ought to support local initiatives the only firm agreement on the State's leading role was about the training of teachers.

The committee did not accept Playfair's diagnosis of Britain's industrial ills, highlighted by his 1867 letter from Paris. Nevertheless it could not be denied that industrialisation on the Continent was progressing extraordinarily rapidly. The committee attributed successful foreign competition, 'where it exists', less to scientific attainment than:

to foreigners' artistic taste, to fashion, to lower wages, and to the absence of trade disputes abroad, and to the greater readiness with which handicraftsmen abroad, in some trades, adapt themselves to new requirements.[54]

The final report echoed the scheme originally promoted by Montagu, himself a member of the committee. It based an analysis of educational provision on the distinct needs of three classes – workmen, foremen, and managers. It called for a thoroughgoing reform of the educational system into an integrated progression. The most urgent need for children destined to be workmen was for elementary education. Lack of this made it impossible for them to take advantage of the adult artisan classes that were available. It also requested a reorganisation of secondary schools to include more science and recommended that some endowed schools which were the subject of the Taunton Commission's investigations ought to be reconstituted as science schools with close links to their local industries. At the top of the hierarchy there would be regional colleges and a national centre in London to serve the various classes of students. The committee argued that the regional college could not run on fees alone but required support from some combination of the State, local interests, endowments and, possibly, the rates. At the same time, it stated firmly that such colleges (and the special schools as well) would have the greatest impact if established in centres of industry 'because the choice of such centres tends to promote the combination of science with practice on the part both of the professors and pupils'.[55]

The emphasis on primary and secondary education would require an expansion of the teaching profession. It was recommended that the education of 'higher' science teachers be encouraged by the granting of science degrees at Oxford and Cambridge as well as by means of encouraging managers of teacher training colleges to promote instruction in science. In other words, the schemes worked out by

Cole in the 1860s had been vindicated. The emphasis of technical education was to be shifted from the artisan to the teacher, the manager and the proprietor. The role of the day science school and above all the tertiary science college was emphasised. This expanded system would require many teachers, who could be trained in the same system. As Playfair had noted fifteen years earlier, and now by analogy with Dublin, the School of Mines in London seemed to be a potentially useful general college of science. For London, it was suggested, further investigation of the complex government scientific establishment was required.

Pure and applied science

The implications for the curriculum of a system such as that suggested by Samuelson were explored by the Society of Arts, whose sub-committee on technical education reported shortly afterwards. The committee, which included the chemists Edward Frankland, Alexander Williamson and David Price, began with the premise that Britain's problem was a lack of scientific knowledge among its managerial classes. That is, the problem was with those who directed works rather than with those who operated them. The committee then defined technical education as the range of competence of its own members: '... general instruction in those sciences, the principles of which are applicable to various employments of life'.[56] However, the committee did not deem formal academic training in the principles of science as sufficient to provide the base on which professional knowledge was to be built. Therefore, it was recommended that after formal academic training there should follow a period of pupilage or apprenticeship. During this period, the student would learn to put into practice the principles that he had acquired. With this orientation the committee suggested that higher education should be devoted to 'the mother sciences', including chemistry. A draft report included detailed proposals for the education of several occupational groups. However, they seemed to be all rather similar. Curricula for chemical manufacturers, metallurgists, miners, and farmers and gardeners were put forward by a specialised sub-committee consisting of Frankland, Williamson, and Price. The education proposed for the four quite different groups differed only slightly, and shared a strong component of the principles of chemistry.[57]

Consequences

That the conclusions of the Samuelson Committee and the Committee of the Society of Arts were complementary was scarcely surprising, since there was a considerable overlap between the witnesses to the one and the committee members of the other. Neither report led directly to action. Nonetheless in their own ways both were important. The Society of Arts report reinforced the continuity between the curricula which had been formulated for chemical education during the 1850s and '60s, the structure of the Royal College of Science in Dublin and the future pattern of technical education on the mainland. The Samuelson Committee made public the institutional plans to bring such science colleges to England and Scotland. When a new Liberal government came to power at the end of 1868 (six months after the Samuelson Committee had reported) the preceding momentum ensured a continuity of concern. At the new government's request, Cole wrote a report of the previous administration's plans for science education. He pointed out that the plans had been published in the Samuelson report and had raised expectations that would need to be satisfied. He was thereby successful in strengthening the hand of his new Minister, William Forster, against the ever suspicious Treasury. Matching grants for organised science schools operating full-time in the day had been recommended by Samuelson, and were brought in by the Liberals.[58]

One casualty of the change of administration was government financial support for the new buildings appeal for Owens College. Nevertheless, the cause of provincial science education was becoming popular in the country. In October 1868 it was endorsed in an *Economist* editorial: an enhanced Owens College was just what the country needed. The Chemistry Department had indeed already proved itself. The success in raising funds was perhaps the best evidence of the attraction of the college's image. £211,000 was raised locally. The splendid building opened in 1871 was not merely a testimonial to civic pride. It was also a demonstration of the college's success in articulating civic ambitions.[59] The general attraction of the vision was demonstrated by the plethora of colleges founded in provincial centres over the next few years. The year of the completion of the Owens building saw the founding on a similar plan of a College of Science in Newcastle, and there were soon a host of other civic colleges: Birmingham and Bristol, 1873; Leeds, 1874; and

Nottingham in 1876. The emergence of these, though anticipated in the 1860s, made all the more significant a new clarification of the nature and function of scientific education.

While Owens College successfully sought to expand, Samuelson and others kept up the pressure for the establishment of a central science school in London on the Dublin model. After his Select Committee reported, Samuelson maintained pressure on the government in Parliament. In July 1869 he used the discussion of the Education Vote to investigate the possibility of combining the Royal School of Mines, the Royal College of Chemistry and the Royal School of Naval Architecture. He pointed to a complex of new buildings then under construction in South Kensington and suggested that this would make the ideal venue for his proposed institution, which would also train teachers.[60] Replying for the government, Forster commented that he wanted to establish a new normal school (school for training teachers) as soon as possible. Significantly he commented, 'It was well-known that the government had the whole system of education under consideration and in considering it, they must of course direct their attention to scientific and technical education.'[61] Those wishing to promote scientific education obviously took the hint.

Since 1860 Henry Cole had been carefully, sometimes discreetly, sometimes publicly, engaged in building a great museum and educational complex in South Kensington. The latter-day survivors of his scheme, the Science Museum and Victoria and Albert Museum, remained anomalous until they were finally removed from the civil service in 1984. All the more remarkable was Cole's success in promoting his scheme against contemporary economic and political philosophies in the era of oft-changing governments during the 1860s and '70s. As his diaries make plain, he took tremendous pleasure in being seen, and sometimes feared, as the successful 'empire builder'. In 1860 a Select Committee on the South Kensington Museum had investigated the way in which the 1851 Commissioners' monies were being spent. An estate in South Kensington near the exhibition site had been purchased and temporary museum buildings erected. The committee approved heartily and recommended that permanent premises be built at government expense.[62] In December 1865 the Committee of Council proposed to the Treasury the funding of a museum complex which would also 'provide efficiently for the growing wants of the Art and Education divisions of the Museum,

their libraries, with the schools of naval architecture, laboratories ...'.
Thus smuggled in, the building, with laboratories, was approved in
early 1866.[63] The next year the fourteenth report of the Science and
Art Department noted that 'the building for the Schools of Naval
Architecture and Science has been begun ... The Schools will
comprise spacious class-rooms, professors' rooms, chemical and
metallurgical laboratories, libraries, specimen museum for objects of
study in metallurgy, chemistry, and naval architecture, together with
a central lecture room adapted to the number of students which the
building will accommodate.'[64] By 30 June 1868 it was reported that the
building was progressing well and that it would be roofed during
1869. Cole suggested that it would have ample room for the School of
Naval Architecture and for a school of chemistry, 'but if other
branches of science are taught, more space will be necessary'.[65] When
the Liberals came to power, some months later, they were rather
surprised to discover the progress that had been made on laboratories
in what had been primarily envisaged by Parliament as a museum.[66]

At Samuelson's request Cole submitted a memorandum about
South Kensington to the Select Committee in July 1868. He proposed
the amalgamation of the School of Mines in Jermyn Street, College of
Chemistry and Royal School of Naval Architecture into a 'college
of science applicable to productive industry'. Suggesting that he had
no specific location in mind, he argued that it would be beneficial for
such a school to be near a 'museum illustrating the progress of science
applied to industry' on the model of the Conservatoire des Arts et
Métiers in Paris. Montagu as questioner prompted Donnelly to say
that this was precisely what had been meant by the central science
school proposed in the ill-fated draft minute of December 1867.[67] In
private a minute drafted in March had already stated, 'Royal School of
Naval Architecture now in operation at South Kensington will also
have to be considered in relation to the contemplated Central College
of Science.'[68] The obliqueness of public reference to South
Kensington was intentional. In his diary Cole noted for 16 May 1868:

To call on Mr Samuelson who wished me to give evidence that a College might
be made of Jermyn Str materials & to avoid deciding where it should be so as
not to excite jealousy of S.K.[69]

There were two kinds of jealousy of which Cole had to beware.
Geologists were apprehensive that the shift of resources away from
Jermyn Street would diminish the geological focus of government

scientific patronage. Indeed, the geologists fought against removal right through the 1870s. Secondly, chemistry teachers in the metropolis, particularly Alexander Williamson at University College, were jealous of the patronage available to the College of Chemistry. Cole's plans ran straight against the interests of both powerful lobbies. His minute to the Samuelson Committee was another way of forcing the issue. Obligingly the report suggested that a possible amalgamation ought to be investigated. However, Cole was unable to keep things private on this occasion. Samuelson himself in his question to Forster in July 1869 first publicly coupled the new South Kensington buildings to the possibility of amalgamating the three government institutions, which could then be located there. In March 1870 Samuelson repeated his question about amalgamating the three institutions and locating them in the new South Kensington buildings.[70] This time Forster had a more specific answer. He reported that a Royal Commission was about to be set up which would investigate the aid from all sources given to scientific societies and which would also investigate how aid could be better given. He promised that Samuelson's problem would come within the range of the commission's enquiries. This was to be the famous Devonshire Commisssion on Scientific Instruction and the Advancement of Science. Linking by its very title research with the teaching of science, the Devonshire Commission showed that there had been a change in emphasis among science educators. Cole noted in his diary that he had had nothing to do with it.[71]

Devonshire Commission: origins

The immediate preliminaries to the setting up of the Devonshire Commission and particularly the role of the British Association in providing a forum for a pressure group seeking State support for science are now well understood.[72] A British Association deputation to the Lord President of the Council in February 1870 requested that both research and education be investigated. Research was a particular interest of that tireless campaigner Alexander Strange, who marshalled the leading scientists of his day. Education had come on to the Association's agenda owing to a request from Williamson at University College, who had wanted to query whether the government had been impartial in supporting institutions of higher education. Possible developments at South Kensington may well have underlain his query. Certainly in testimony to the commission he expressed his concern about unfair competition from State-supported institutions, meaning presumably the Royal College of Chemistry.[73]

The commission established by Forster was chaired by the prestigious Duke of Devonshire, an aristocratic believer in science who gave his family name to the new physics laboratory at Cambridge. On the commission sat a variety of leading scientific men, including Huxley. Many of its conclusions, such as the need for a governmental Department of Science, resonated a century later. The commission's call for government patronage for research seems to have been best remembered.[74] However, the first of the commission's eight reports issued over the next five years dealt with the Royal School of Mines, the Royal College of Chemistry and the Royal School of Naval Architecture. The second report dealt with government support for science teaching at elementary and secondary levels (Science and Art Department examinations) and with the training of teachers. Thus all the questions that had remained unsettled from the 1860s were at the very top of the agenda and were now reinvestigated in the light of the passage of the Forster Act of 1870. The third report considered the traditional universities, while the fourth dealt with government sponsored museums and the fifth investigated other universities and colleges in England and Ireland. The sixth report returned to other unfinished questions from the 1860s – the teaching of science in the private sector at public and endowed schools. The seventh commented on the University of London and the Scottish universities. It was left to the eighth report to investigate what support the government did and might give directly to research.[75]

The discussion on the future of the South Kensington laboratories highlighted the fundamental discussions about the future of science education. The Samuelson Committee had suggested a dual role of scientific training for applied scientists and for teachers. The Devonshire Commission shifted the balance to the teachers. Thereby pure science, as the professional education of teachers and as the universal basis of education for all technical professions, was defined as the principal focus of the curriculum. The diverse pattern of political and institutional issues involved can be seen in the thrust and counter-thrust of the witnesses' interviews.

Devonshire Commission: testimony

Again Henry Cole was the the first witness to testify, in mid-June 1870. This time he had to defend his plans from attacks on many fronts. He had already been challenged by the Treasury, which had begun to scrutinise rather more stringently the operations of the Science and Art Department. Cole had begun to come under censure for his handling of financial matters. It was not only a question of whether the books were balanced, but whether the department was taking initiatives independently, before they had full Treasury approval. The Secretary of the Treasury's membership of the Committee of Council of Education, though intended to prevent such evasion of control, was an inadequate precaution, since decisions were often taken by a small executive, and the full body seldom met. It was in a context of fiscal and policy surveillance that Cole gave his testimony.[76] He was also challenged by the scientific establishment who had forced the setting up of the commission. Cole was made very aware of the private colleges' concern at competition from government institutions. Ten days before his testimony to the commission he had been visited by Lockyer, editor of the new and influential journal *Nature* and Secretary to the Commission, who had impressed upon him the need to appease the major London colleges, University College and King's College.[77]

The questioning was opened by the chairman, who simply asked Cole to describe the Science and Art Department system and to relate its history. Samuelson, as a member of the commission, was the second questioner and immediately began to steer Cole along familiar lines.[78] Was the department satisfied with the degree of competence of its teachers? 'Yes', said Cole, but he immediately qualified his assent by emphasising that it was only in relation to the existing system of limiting scientific instruction to part-time evening classes. For the sort of system of advanced scientific instruction that Britain ought to have, the competence of teachers was quite inadequate. When asked how he would go about obtaining a corps of better qualified science teachers, Cole harked back to the memorandum he had submitted to the Samuelson Committee recommending the amalgamation of the Royal School of Mines, the Royal College of Chemistry and the Royal School of Naval Architecture into a 'Metropolitan College of Science, the main object of which was that it should be a training school for teachers'. Where should it be? 'Of course', said Cole, 'in my mind

there is no place in the world like South Kensington, but I should say
that they could be most economically placed at South Kensington.'
Then Samuelson turned to the question of buildings. Could such an
institution actually be housed in South Kensington? Straightfacedly,
Cole explained that there were no buildings:

specifically for a school of science. When it was resolved to build permanent
premises for the School of Naval Architecture and Science, and to get them
out of their present wooden sheds, provision was made in the upper part of the
building for a laboratory of considerable size, in the view of those [teachers']
scientific examinations, as well as bearing in mind the knowledge that Jermyn
Street was overcrowded with those laboratory students as well as the College
of Chemistry; but the building at Kensington was not made with the view of
superseding the School of Mines in Jermyn Street. It would conduce much
towards it, and with additions might be turned into a general school of science
with developments into Naval Architecture and Marine Engineering and
anything else that might be desired.[79]

In reply to further probing, Cole reckoned that the site could be made
fit for such purposes within a year. Six months later he and Donnelly
provided specific costings: another £45,000 would be needed to
complete the buildings for a proper training school.[80]

The questioning then passed to William Sharpey, Professor of
Physiology at University College. He probed the issue of State
competition with private enterprise institutions. Cole stressed that his
scheme was for training science teachers. A basic knowledge of
science could be obtained in many institutions in England and indeed
in London, but the 'acquisition of general knowledge, and the power
of efficiently imparting it, are two different things'. The latter
required special instruction in a special institution.[81] A particular
concern for Sharpey was the College of Chemistry, and Cole could not
avoid saying that in this case the government did seem to be
competing with other institutions, though he pointed out that the
situation had arisen inadvertently, as it were.[82] The questions of
Huxley, Professor of Natural History at the School of Mines and close
associate of Cole, tried to help Cole retrieve the situation. He
encouraged Cole to explore ways in which the State might sensibly
sponsor science instruction without being competitive. Cole came
down in favour of widening State help for science teachers without
State competition to private organisations. He favoured converting
current State endowments, such as Regius Chairs at the various
universities, into scholarships for which students could compete by

examination and which they could then use at any institution of their choice.[83] To a subsequent questioner he advocated additionally building and apparatus grants.[84]

Cole's position was obviously very difficult as he tried to tread a safe course amongst the various questioners. In reply to Sir James Kay-Shuttleworth, pioneer of elementary education and teacher training in England, Cole suggested that one year might be sufficient time at the central science school to train an already certificated teacher in science. On the other hand, lest this provide a reason for doing away with the Royal School of Mines or the Royal School of Naval Architecture, he argued that their work could form the core of a student's third year, along with other choices of applied topics on the Dublin model.[85] Following this line, W. A. Miller, Professor of Chemistry at King's College, enquired whether the new central school would would then be limited only to training science teachers and Cole had to answer in the negative. How then was competition to be avoided, asked Miller? Cole could not deny that a certain amount would be inevitable.[86]

The matter of competition was not the only difficult area for Cole. Kay-Shuttleworth probed the area of just whom the new teachers would teach. The upshot was that Cole had to confess that he hoped for a new system of scientific day schools around the country, although the take-up of the Science and Art Department's building grant for such purposes had as yet been insignificant.[87] He answered a similar question from the chairman by analogy with the Art Training School - it had been impossible to outrun public demand for teachers. Cole hoped for a secondary school in every centre with a population of 10,000 or more. Quite apart from science schools, each secondary school ought to employ at least one or two science teachers.[88] So the whole of Cole's scheme was shown to be dependent upon the reorganisation of the education system that was being projected and carried out by William Forster.

The provision of laboratory training for science teachers became a new element in the arguments about a central science college. Hitherto neither teachers nor pupils had routinely had laboratory experience. Edward Frankland, by now Professor of Chemistry at the Royal School of Mines, described to the Devonshire Commission the week-long summer laboratory course he had given to prospective chemistry teachers in the previous summer. He had had twenty-six pupils then and he wanted to increase that number to 120 the next year. At forty per week, over three weeks, such numbers would put a great strain on Frankland's facilities.[89] In 1875 he would publish a famous set of basic experiments, *How to Teach Chemistry*, for Science and Art Department teachers.[90] However, it was not until 1878 that examinations in practical chemistry were introduced into the Science and Art Department curriculum.

Frankland testified twice, as examiner to the Science and Art Department and as Professor of Chemistry, each time with a different emphasis. In his capacity as examiner he mentioned that he was aware that there was a demand for the systematic training of Science and Art Department teachers, but, rather disengenuously, that he was not aware of any moves to give more laboratory training to prospective teachers. However, he rather gave the game away as regards the buildings at South Kensington. When asked by Samuelson whether he was familiar with them, he replied that he had in fact visited them a year or two before. 'They were projected principally, I am informed, with the view of affording such instruction to teachers who would afterwards be distributed throughout the country.'[91] Indeed, Frankland was aware that several separate laboratories were intended for different grades of expertise. In his testimony, for the first time the full scheme considered for South Kensington was revealed:

The plans of the buildings were arranged for three or four separate laboratories; there was to be a laboratory of beginners and a laboratory for more advanced students, one for experimental research, and one for the technical applications of chemistry. These separate laboratories are absolutely essential for efficient instruction, and several sets of rooms for distinct kinds of operation are required for each of these laboratories.[92]

When pressed by Samuelson on whether, given limited funds, he would prefer support for elementary or for advanced education, Frankland was in danger of having to choose between the goals of training teachers (demand for whom depended on elementary classes) and the higher education of manufacturers. Frankland unequivocally responded in favour of advanced education. Huxley, as Frankland's next questioner and co-examiner for the Science and Art Department, managed to manoeuvre Frankland into retrieving the situation somewhat, by getting him to say that it would be regrettable if any diminution of support for elementary education led to fewer students coming forward for advanced training. Not to be denied the last word, Samuelson then asked Frankland whether he thought the rudimentary scientific training of workmen or the advanced training of managers was likely to be more beneficial to British industry. Frankland admitted that, given only one possibility, the instruction of managers would be preferred.[93]

Some eight months later, Frankland testified again to the Devonshire Commission, this time as Professor of Chemistry at the Royal College of Chemistry. In contrast to the information he had provided to the Samuelson Committee three years before, he now argued that his laboratory was too small for his ordinary students, let alone for the number of teachers who might come in the summer. 'Several important branches of chemistry either cannot be taught at all, or can only be taught very imperfectly for want of more space.' Gas analysis was hardly taught, spectrum analysis - the most important modern tool - was totally ignored; inorganic quantitative analysis was taught imperfectly for want of a balance room; and the private laboratory had been given over to more advanced students, thereby eliminating the possibility of doing research at the college at all. He could not estimate just how big an expansion might be warranted, but he was always having to turn students away for lack of space.[94] He revealed a little more about the proposed laboratories in South Kensington. Hofmann had been consulted about them and after taking over Hofmann's professorship Frankland too had been requested to look at the plans. He certainly favoured the move to South Kensington and admitted that if the new buildings were given over entirely to chemistry, he would have facilities comparable to the very best on the Continent.[95]

When challenged about the unfairness of State support for certain teaching institutions, Frankland was less definite. To one questioner he suggested that a State-supported school would probably boost science to such an extent that it would benefit all schools.[96] To Samuelson he argued that the college concentrated much more on the applications of chemistry to manufactures than any other institution. In any case, the proposed school at South Kensington was to be primarily for teachers.[97] Huxley's pointed questioning enabled Frankland to extrapolate from the recent rapid growth of students: the number of laboratory students over the preceding twenty-five years in London had multiplied tenfold and that number was likely to continue increasing. Therefore there was plenty of scope for more science teaching, by private and by State institutions alike.[98]

As for research, Frankland was convinced that special State grants and State institutions for research as proposed by the British Association were not the answer to the problem of increasing British output. What was required was more recognition for research, preferably through university degrees. Once research was given what he saw as appropriate status, then increased grant support might be usefully sought. Until then it would be sterile.[99] The vision of Frankland, as chemistry professor at the Royal College of Chemistry, was clearly compatible with that of Cole, the administrative entrepreneur. An enhanced system of colleges would train teachers and others in advanced science and would be the basis of research.[100]

Speaking for privately funded institutions, Williamson, recently appointed dean of the new science faculty at University College London, agreed with many of Frankland's general ideas on the teaching of science and the importance of research. He was, however, against specially favoured State-supported institutions, which could not but compete with struggling private institutions. Students would go there with a view to obtaining government appointments later. He was particularly perturbed that the South Kensington buildings had been undertaken without consulting the scientific community outside government institutions.

Williamson's evidence to the Devonshire Commission emphasised the philosophy articulated in his recent address 'A Plea for Pure Science'.[101] Adamant that academe was no place to learn the applications of science, Williamson argued that that was best done in the 'real life' situation of the works after a solid foundation of the study of scientific principles. The one vocation for which colleges could offer training was teaching, for the university actually was a 'works' as far as teaching went. He favoured a professorial pupil-teacher system, 'and that is really the only natural system of a normal school'. Indeed, many of his former pupils had become teachers. The obvious conclusion was that no special training institutions were needed either for future industrialists or for future teachers. The university already did the job.[102] Explicitly Williamson objected to what he called 'technical schools'. By these he did not mean the actual School of Mines or the College of Chemistry as they were currently run, but rather French-style polytechnics such as the School of Naval Architecture.

Whereas I should wish that the Government should acknowledge by its acts the importance of pure science, their establishment of technical schools in opposition to schools of science, does seem to constitute a recommendation to something else than the study of pure science, and to favour a popular prejudice, which gives us a deal of trouble, that there is a royal road to efficiency.[103]

As regards research, Williamson was equally firm. This too was to be the function of the academic. It was vital that training and research be combined.[104]

The complementary arguments of the Science and Art Department and the professoriate seem to have convinced the Devonshire Commission. In its first report, published in March 1871, the commission suggested that the School of Mines and the Royal College of Chemistry already constituted 'one school of Pure and Applied Science'. It was their organisation that was at fault. A Chair of mathematics was required, laboratories for physics and biology, and better facilities for the College of Chemistry were also needed to make the college efficient. The commission recommended the proposed merger of the three government science teaching institutions to form a science college for teachers in the new building rising in South Kensington.[105] This recommendation was confirmed by the later reports covering teacher training and science teaching in schools.[106]

Thus the plans worked out during the 1860s by the Science and Art Department had been vindicated again. The government had been urged to encourage a central science college with the central function of training teachers rather than as a polytechnic. There were important consequences for chemistry as an academic subject. In the polytechnic-type course where chemistry was a means to a practical end, it would play a minor role commensurate with its direct utility. However, in the science college chemistry, underpinning many other areas of endeavour, was a subject with few equals. Whatever subject one specialised in, the study of chemistry was a compulsory subject for the London BSc. Moreover chemistry was unchallenged as the favourite subject for London BScs for the rest of the century. Ironically chemistry, which had always been championed as the most universally applicable of the sciences, thrived especially as a pure science taught separately from its applications.

Chapter Six:
Epilogue

The categories of pure and applied science articulated by the Devonshire Commission infused subsequent discussions. They permeated the structure of a variety of new institutions formed in the late nineteenth century. Chemistry was reformulated in novel terms, not only in education, but also in the relations between sectors of the chemical community. The formal coalition represented by the Chemical Society was found wanting, and separate institutions were set up for professionals and manufacturers. Nonetheless, because their leadership was academic even they asserted the primacy of pure science. This epilogue will briefly sketch the main lines of development up to the First World War, showing how intellectual categories acquired institutional reality.

Though few of the specific recommendations of the Devonshire Commission were immediately implemented, the arguments remained telling and almost all were realised during the following century. More immediately the commission's reports created a new intellectual environment for scientific and technical education in Britain. They laid the groundwork for the subsequent construction of a binary system of education in which universities and technical colleges developed as two different kinds of organisation. The commission established the principles on which the 'academic' side was to be built and established a hierarchy of knowledge. At the head was academic science. Firmly in control of what constituted appropriate knowledge were academics, rather than representatives of the practising community.

Meanwhile the polytechnic idea remained potent.[1] A vigorous movement for the establishment of a 'technical university' continued to be active through the 1870s and 1880s. Indeed, through its efforts, a system for technical education was codified in terms almost parallel to that defined for scientific education by the Devonshire Commission. Examinations, paid-by-results, regional technological colleges, and a central technical institution in South Kensington for training teachers and highfliers were all proposed and one by one enacted. The movement even had its own Royal Commission on Technical Instruction, which sat from 1882 to 1884 under Samuelson's experienced chairmanship. This movement effectively represented the establishment of the basis of the second sector of the binary system. In 1889 the Technical Instruction Act authorised local authorities to spend centrally collected funds on technical education. Effectively, the scientific and the technical were divorced, the former aimed at the middle classes, the latter at improving artisans.

The general thrust of the Devonshire Commission was indicated by its recommendation for South Kensington. The arguments over the implementation of those recommendations show how the science college came to be the dominant model. The controversy over the proposed amalgamation of the College of Chemistry, the Royal School of Mines and the Royal School of Naval Architecture on the South Kensington site needs to be seen partly in the light of the parallel debate about a technical university.

South Kensington: an 1880 view

It was not until 1881, some ten years after the initial recommendation, that all the participants agreed and the Treasury sanctioned the move.[2] For the 1881-82 academic year a new institution called the 'Normal School of Science and Royal School of Mines' opened in the now not-so-new buildings in South Kensington. In order to argue the move through the Treasury, Donnelly prepared a position paper late in 1880 summarising the evolution of the School of Mines and the South Kensington site to date. He argued that it was only sensible to make official what was already the case. South Kensington, whatever its rhetoric and intentions, had already evolved into a science college and an important training institution for science teachers. Donnelly's analysis of the reality of the developments was that:

the school instead of developing as was contemplated in 1853 into a Normal or Training School of General Science has become, under the name of School of Mines, a place where a few persons connected with the mining industry are educated in science, and where a much larger number attracted by the eminence of the professors, and disregarding the name of the school, either take up special subjects, or go through its curriculum in order to follow industrial pursuits in no way connected with mining.[3]

In addition, while the school had not officially become a training school for science teachers, there was no doubt in Donnelly's mind that:

Such a school has come into existence through force of circumstances, and in spite of many difficulties, as an outgrowth of it [the RSM] – an outgrowth which is larger and more important than the parent stock.[4]

Donnelly backed up his argument by surveying the students' career goals in that year. Of 200 students enrolled, fifty-three were formally teachers in training, thirty intended to be mining engineers, and another twelve envisaged some sort of metallurgical career either as assayers or as ironmasters. He noted that a large number intended to be analytical chemists or works chemists, while a few were going into a range of industries such as glass, paper and brewing, and others into the traditional professions. Such was the output of the science college, even imperfectly realised.[5]

Devonshire Commission: immediate response

The commission's recommendation had not met with ready acceptance, falling foul of both the 'geological' party at the Royal School of Mines, and of the Treasury officials. The 'geologicals' baulked because the formation of the new science college in South Kensington would involve the violation of a long-established principle – the close linking of the Royal School of Mines and the Geological Survey. Furthermore, maintaining the view of the school as a training centre for a particular industry in opposition to those pressing for a science college, the 'geologicals' objected to the commission's allegations that teaching at Jermyn Street was inefficient. On the contrary, they argued, it was well organised as a mining school, it was only when unsolicited wider objectives were overlaid upon it that the charge of inefficiency might be justified. A lack of a biological laboratory was hardly a problem for a mining college. They also suggested that while there was justification for government finance of a specialist body such as a mining school, which would offer no competition to existing institutions, the proposed science college would infringe the principle of non-competition from State-funded institutions.[6]

The Devonshire Commissioners' response conceded none of these arguments. Most pointedly, the responding sub-committee noted that the question of loss of identity was hardly a serious matter when, over the preceding eighteen years, only some four students per annum who had been examined in mining or mineralogy entered relevant professions.

> ... and if, notwithstanding the eminence of the Professors, the mining interest has been very slow to avail itself of the advantages afforded by the School, this may be owing in a great degree to the imperfection of its organisation, and to the incompleteness of the preparation which it supplies for the professions of mining and metallurgy ...[7]

More grave in the commissioners' eyes than the lack of biology, so far as mining was concerned, was the absence of instruction in mathematics. That lacuna would be filled from the School of Naval Architecture.

So the commission was unmoved. In February 1872, by which time Murchison had died, it made specific recommendations about the organisation of the new college and even the disposition of the space in the South Kensington building. Chemistry did very well indeed in their proposals and mining was rather diminished:

Basement – Physics, Metallurgy, Chemistry
Ground – School of Naval Architecture, General Lecture Room, Mathematics, Applied Mechanics, etc.
First – Physics and Chemistry
Second – Chemistry
Third – Biology, Mineralogy, Mining, Geology, Physics, Chemistry (open-air work)[8]

At last the extent of the conspiracy over the new buildings was revealed: the School of Naval Architecture was to occupy only one of the four floors, the rest was devoted to the general science college and a large proportion of that to chemistry.

The geological faction at the Royal School of Mines was not unsupported in its opposition to the establishment of a science college at South Kensington. The matter was taken up actively in the press. In a series of splendidly polemical articles the popular journal, *The Engineer*, pitched into the first report of the Devonshire Commission. Not only was the whole report a piece of 'malicious jobbery' on behalf of the professors and an 'unveiling to the public in decorous style of the absorptive longings of Mr Cole C.B.', the concept of the science school would be entirely inconsistent with practical education. The entirely useful, though undercapitalised, School of Mines would be replaced by some 'nondescript "Science School"' teaching elementary science out of which the Schools of Mines and Naval Architecture would be expected to grow like 'toadstools on rotten wood'.[9]

In the event, according to Huxley (Professor of Natural History at the School of Mines), he, Frankland, and the professor of physics refused to go on teaching unless a transfer to South Kensington were made. The geological faction had to capitulate and agree that it was impossible to find sufficient accommodation at Jermyn Street. In July 1872 the Council of Professors recommended that physics, chemistry and natural history be transferred to the new facilities in South Kensington. Clearly it was a compromise, with the non-geological 'academic' faction getting its new premises, but with the old administrative structure being retained. The Museum, the Survey and the School of Mines were still closely linked; instruction in the 'more specialist' subjects of mining, mineralogy, metallurgy and palaeontology remained in Jermyn Street. The transfer was effected quickly, with students starting at South Kensington in time for the 1872-73 academic year and the naval school moved out.[10]

Considering the question in 1881, the Treasury was happy with the designation of South Kensington as a Normal School. This limitation would both absolve the government from charges of competition with private institutions and forestall arguments about other ways in which it might better spend its money on science. At the same time it was reluctant to sanction the elimination of 'a strictly technical and professional school of mining knowledge'. Murchison, and his lone surviving standard bearer at the School of Mines, Warington Smyth, the Professor of Mining and Mineralogy, had had a point: the power and prosperity of Britain rested 'in an especial degree' upon mining. Whatever intentions the originators of the school had had in the early 1850s, it was clear that the government of the day had felt a primary commitment to the application of science to the mineral riches of Britain and its colonies. The current Treasury renewed this commitment: the School of Mines ought not to be subsumed by a general science school, but deserved independent consideration:

to preserve its direct technical character. Preceding governments appear to have recognised this particular industry as one which more than most others

called for scientific knowledge on the part of those engaged in it, was of an
importance which justified public assistance to promote it ...[11]

The scientific departments of the School of Mines formally became
the Normal School of Science, but the Royal School of Mines
maintained a separate identity and specialist instruction in mining.
Rather than a single integrated institution, the proposed great central
school, as its title shows, in fact comprised two juxtaposed
institutions. The government had still shown itself willing only to
support specific vocational goals. Even the term 'Normal School' was
analagous to the French 'Ecole Normale'. Clearly this was a
compromise model: it was not quite a science college nor a
polytechnic. It was the former in that it offered a general basic
education in the principles of the fundamental sciences while it
proffered training for teachers. It was the latter in that it offered
technical training in an applied field. However, primacy was given to
pure science, and applied science under the heading of mining was
clearly of secondary importance. Although students passed regularly
between the two segments of the institution, the scientific and the
technical were in fact institutionally distinct. The distinction was
emphasised even further when in 1890 the title of the combined
institution was changed again to 'Royal College of Science and Royal
School of Mines'. This was nearer to the ideal of a central science
college, but still afforded the same emphasis to the separate technical
school. The geography of the institution was also simplified in 1890
when Warington Smyth died, and the last remaining courses at
Jermyn Street were transferred to South Kensington.[12]

Thus the shift to a science college from a polytechnic was no simple
ideological decision. Giving strength to a pure-applied science
dichotomy, it had reflected the government's beliefs about the
legitimate uses of public money, the strength of the geological lobby,
and the power of the academics. Chemistry, which had always been a
marginal part of the old School of Mines, became a central part of the
new college.

Science and Art Department examinations

The other wing of the Science and Art Department's activities, the
examination system, maintained a similar separation between science
and practice throughout its existence until 1897, when the payment

by results system was ended. Increasingly study was undertaken in daytime 'organised science schools'. The system grew steadily, so that in its final year there were almost 200,000 students with just over 100,000 examinees. The numbers taking examinations in inorganic chemistry increased from 2,700 in 1870 to a peak of almost 24,000 in 1895.[13] Though chemistry had by then dropped in the league table of examination subjects, it still attracted a significant percentage of total examinees, fluctuating between 12 and 21 per cent from 1870 to 1895. An 1890s directory lists some 700 Science and Art Department teachers of chemistry throughout the land.[14] It has been argued that the success of the system was not only due to the inherent attractions of the examinations or their subject matter. The Science and Art Department examinations filled a gap in the educational structure established by the Forster Act. The examination scheme became a *de facto* secondary education system in an era when there was no regular State provision for it.[15]

The Science and Art Department examinations retained their purity of subject coverage, while complementary 'practical' education was taken up by the Society of Arts technological examination from 1873 and by the City and Guilds examinations six years later. Thus there was an incipient binary system on the level of part-time evening study as well as in the full-time tertiary sector. The Society of Arts examinations in technological subjects were not quite the high-level professional certificates forming the pinnacle of the training of prospective managers envisaged in 1868. Rather, following a suggestion of Donnelly, they were designed for secondary-level part-time evening students to complement the Science and Art Department subjects in evening schools.[16] The combined system would offer three stages of examination leading to qualifications in particular branches of technology. The first stage, already covered by the Science and Art Department, was a general examination in the principles of the sciences relevant to the student's specialist or applied field. The second stage to be tested by a new Society of Arts examination would be concerned with the principles of manufacture. It was hoped that a third stage, the practical skills of manufacture, would eventually be picked up by a series of professional institutions.

From 1879 the Society's examinations were taken over by the City and Guilds of London Institute for the Advancement of Technical Education. For under the threat of losing their ancient endowments, the old and wealthy City of London livery companies had directed their attention to technical education. The target audience was the manufacturing population – workmen, foremen and managers. They were to be offered subjects much more practical than the Science and Art Department offerings. For example, the City and Guilds counterpart of the Science and Art Department examination on inorganic chemistry was alkali manufacture. It was hoped that these examinations would be regarded as certificates of proficiency which would enable operatives to improve their position. Although the subjects covered a range from handicrafts to high technology, the numbers of examinees remained considerably less for the rest of the century than in the Science and Art Department system – 816 examinees in 1880 and 4,105 in 1900. The split between the two examination systems, one concentrating on principles and the other on applications, represented a powerful institutional form of the pure-applied division.

The City and Guilds also revived the idea of a large technical university – in effect a polytechnic – to be located in South Kensington. First mooted at the time of the first Devonshire Commission report, this idea was actively pursued from 1879 onwards, and the Central Technical College opened opposite Cole's science school in 1884. It complemented its neighbour: the first year of the intended three-year course was to be devoted to general science and the second and third years to engineering, physical or chemical subjects. The college was seen as the culmination of an education which might have begun at the level of local City and Guilds examinations and progressed through a local technical college. It was anticipated that the students would include potential teachers of technology, architects, engineers and manufacturers as well as mature students wishing to retrain.[17]

In practice the working out of these schemes was far from clear-cut. Just as the Royal College of Science was for many years a messy compromise between polytechnic and science college, so at the Central Technical College the first Professor of Chemistry was a pupil of Frankland, H. E. Armstrong. Ironically for an institution established for its practicallity, he was specially well known for his faith in the research method. By contrast at University College, where Williamson had preached 'A Plea for Pure Science', the eminent brewing chemist Charles Graham was appointed Professor of Chemical Technology in 1877. Graham disagreed profoundly and publicly with Williamson over the role of pure chemistry in relation to industry. He felt that technology itself should be taught.[18] Despite his belief in chemical technology, his successor in 1888 was appointed to a lectureship in applied chemistry, a post that only endured another nine years till that last remnant of Graham's attempt petered out.

In the provincial centres there was also no precise demarcation between institutions. It had been a conclusion of the 1870s debates that while provincial education should have national encouragement, much would be left to local initiative. Thus the system that developed nationally was complex and the case of chemistry was no exception. The Devonshire and Samuelson Commissions and the structure of the national examination systems were very influential. Moreover the personal influence of Roscoe, Sanderson has argued, as 'a founder of modern chemical education in this country' is impossible to exaggerate.[19] At the same time local pressures and concerns modulated local realities. In Manchester two great institutions emerged: Owens, home of the first professorship of organic chemistry, though it did start a technological chemistry course in 1883, allowed technical specialisation only in the third year, and did not offer engineering to the students.[20] Meanwhile the Manchester Mechanics' Institution evolved into Manchester Technical School in 1882. There, much more than in Owens, individual trades were taught. The three laboratories for chemistry were devoted to pure chemistry; dyeing, bleaching and printing; and to metallurgy. The results of City and Guilds examinations of 1890 showed that the school was the best of the provincial institutes.[21] The lectures there of George Davis were the basis of his seminal textbook on chemical engineering. Elsewhere the divisions between university and technical institute were not so clear cut. Yorkshire College in Leeds created Departments of Tinctorial Chemistry, Leather Chemistry and Gas Engineering, and Birmingham University created the British School of Malting and Brewing. But even there the diversity of possible applications of chemistry encouraged the formation of an all-purpose central school of pure chemistry.[22]

The Society of Arts plan that there should be a third stage of examination to determine technological proficiency run by professional institutions was never actually implemented. Nevertheless, new central institutions attempted to impose professional standards on the apparent confusion.[23] In the case of chemistry, discussions were particularly intense as the automatic coalition between academics, manufacturers and other members of the practising sector ceased to be tenable. New legislation demanding certified chemical expertise to deal with a variety of newly significant social problems caused particular anxiety within the chemical community. The Pharmacy Act of 1868 required that pharmacists and chemists and druggists should have specific qualifications and raised again the not insignificant problem of the use of the term 'chemist' by pharmacists. Similarly, the Sale of Food and Drugs Act of 1872 provided for the creation of Public Analyst posts and recommended that their incumbents have relevant medical, chemical and microscopical knowledge. The Public Health Act of 1875 stipulated that Medical Officers of Health should hold a special diploma based on the study of a range of subjects, including chemical analysis, relevant to public health problems. These statutory requirements raised chemistry's by then endemic problem: which sector – academic or practising – was to be regarded as the custodian of this particular form of knowledge? Were the academics, or the practitioners, the 'true' professionals? How were they to be trained? Chemists repeated the discussions about pure and applied science, polytechnics and colleges. Did the principles of science or intimate experience in the field constitute expert knowledge? The academics, while recognising the essential role of practical experience for complete training, nonetheless asserted the primacy of academic knowledge. Because of their primacy in pure science they saw themselves as the guardians of practising chemistry. Practitioners, by contrast, relegated pure science to the category of interesting background knowledge, which gave a bit of polish, but was hardly essential. The establishment of the Institute of Chemistry in 1877 demonstrated this realignment of the chemical community with academics firmly at the head of the hierarchy.

Analysis was at the centre of the debate, for all the chemical skills required for the new statutory positions were in essence analytical. However, these skills had to be applied in very specific contexts rather than in the abstract manner of the academic laboratory. An analysis of butter or milk with a view to pronouncing on its quality was a very different problem from elementary analysis in the laboratory. The implementation of the new legislation required new standards without precedent.[24] That the academic community should feel it had a vested interest in this area is hardly surprising. As has been seen, analysis was the cornerstone of the academic currriculum. If rhetorical claims about the general utility of analysis were to be upheld, then the new legislation was but creating new fields of endeavour. Here was a whole new series of posts being created just as academic chemistry departments around the country were expanding. We have already noted the expansion at Edinburgh, Owens and South Kensington as well as the appearance during the 1870s of a number of the new university colleges. It was very much in the interests of the academics that the training which they offered should be seen as providing the qualifying expertise.

At the same time, segments of the practising community, particularly the professionals, argued strongly that laboratory and actual circumstances were so distinct that specific vocational training needed to be the mainstay of the new qualifications. Within the Chemical Society some members wished to set up specific qualifications for membership so that the Society might itself validate the new pracititioners. This only put the problem at one remove, because there was still the issue of what should constitute the new qualification. Disputes broke out within the Society when some members began to exercise their right to blackball candidates whom they deemed unsuitable – in this case unqualified – for membership. At the same time, some of the Society's members began to use their membership as a *de facto* qualification.[25]

It was not only a matter of what training should be given, or how it should be acquired (in the classroom and academic laboratory or in the field or by some combination of the two), but how such training was to be validated. Medical men had their degrees and their examinations for the various professional bodies. Other than the three-year degree of the University of London, there was no specific examination available to test prospective chemists.

In 1874 a group of practising public analysts met to form the Society of Public Analysts, which aimed to create analytical standards for use by the profession. Adamant about the irrelevance of a purely academic training for their calling, the Society was irritated by the suggestion made before the Select Committee on the Pure Foods and Drugs Act that public analysts might well be examined by Frankland and his colleagues at South Kensington.[26] However, the Society did not set up its own qualifications.

Academics and the more status-conscious academically trained professionals were mainly responsible for the establishment of the Institute of Chemistry in 1877. It was a 'professional' organisation concerned with establishing and validating the qualifications of chemists for the new posts and any others. Edward Frankland was the first president and it is hardly surprising that the Institute took a markedly academic turn from the start. It was necessary immediately to face two issues. Firstly, given that there were as yet no recognised qualifications for chemists and that membership would in itself be a qualification, who should be admitted to the new Institute? Should all comers be welcome, or should there be academic criteria for admission? Failing that, should a specified number of years of practical experience be required? The academic party held sway after an initial period of grace during which new members were admitted on their merits as judged by the new Institute's council. The core of the training of the new professional chemist was to be a three years' systematic course in theoretical and applied chemistry, physics and mathematics. The emphasis was on a sound education in the general principles of the science rather than on specific vocational training. In such an uncharted field with so many varied applications, ran the by then familiar argument, a general preparation would be the most suitable grounding for work in any area. In this way, academics within the Institute cast their own training as the ideal qualification for the subject. In the initial regulations, the three years' general training only qualified the applicant for the Associateship examination of the Institute. Evidence of a further three years in practice (which was specifically construed to include teaching and research) was required before the Associate could proceed to the Fellowship. Though academics could still lay claim to professional status as practitioners, the arrangements bowed to the practising interest by making vocational experience a prerequisite for professional validation.

Later the academic influence on the Institute was strengthened further. After 1883 the three years' systematic study had to be done at certain specific institutions, and had to be validated by passing the Institute's examination or by obtaining a degree. The rules came to mean full-time study in the day, for the new requirements specifically excluded the Science and Art Department examinations in chemistry as exemptions from the Institute's examinations. By this move the Institute effectively eliminated a whole range of institutions which might have been attractive to potential practising chemists who were already employed during the day. It goes without saying that the institute did not recognise the City and Guilds technological examinations either. Thus the Institute itself came to embody a self-styled elite of practice, while it excluded major categories of practitioners. The academic emphasis both in curriculum and in the specification of full-time study continued until well after the Second World War.

The Institute concentrated initially on analytical and consulting chemists in private or public practice. It was this group for which qualifications were an urgent matter. However, in industry too the principal role of chemists was analytical at this time, and the Institute argued that the general training which it proposed would serve the industrial chemist as well. But in the last quarter of the nineteenth century there was no pressing reason for chemists destined for industry to seek formal qualifications, let alone a systematic three years' course. It has been calculated that in 1902 only a quarter of the country's graduate chemists were employed in industry.[27] Nevertheless, a decade after its analytical regulations, the Institute introduced its famous branch examinations in a number of areas of applied chemistry. These examinations could be taken after the three years' systematic academic course. Pure science as taught at university level was to be the basis of even industrial chemical knowledge.

The Institute maintained a consistent attitude towards part-time study and the value of strictly academic training. It was the government which finally brought about the involvement of the professional bodies in the business of validating qualifications earned part-time as a part of the reconstruction programme after the first World War. From the end of the Science and Art Department system at the turn of the century every technical school and college had run its own courses and given its own awards. The National Certificate scheme was designed to promote uniformity and to encourage systematic study of programmes rather than hit-or-miss attendence at unstructured groups of courses. The certificates were to be awarded to part-time students who systematically followed courses and passed examinations in approved groups of subjects at non-university institutions. Syllabuses were to be designed locally in order to meet the needs of local industries, but they had to conform to broad national guidelines. The scheme was specificially meant to attract part-time students working in industry, and there was to be special emphasis on broad practical training. There were to be two levels of award – the lower (eventually called the Ordinary National Certificate ONC) required three years of part-time study. The upper or Higher National Certificate (HNC) was originally meant to be equivalent in standard to the Final BSc at pass degree level.[28]

The role of the Institute, in common with the other professional associations, was to act jointly with the Board of Education as external assessor of local facilities, syllabuses and examinations in its own discipline. It had been the intention of the Board of Education to foster more practical vocational training rather than general professional learning. However, the still powerful academic party at the Institute of Chemistry insisted that the core of the National Certificates in chemistry should call for broad training in the principles of chemistry, physics and mathematics. Any applied studies included in the later years of the course had to be based on a general investigation of the principles of the particular industry involved, and were to exclude works techniques. Ironically even though the Institute was successful in asserting academic authority here, it refused to allow the certificates to be viewed as professional qualifications in their own right, or to accept the technical college course for them as satisfactory training leading to its own qualifications. The certificate programmes, it was explained, did not stress sufficiently a preliminary general school education, offered courses that were too specialist (meaning too technical), and did not achieve a high enough standard in the basic sciences. Thus, until a change of policy in 1956, professional status was not granted to those educated solely on the technical side of the 'binary' system.[29]

Just as the Institute was deciding on stringently academic entry requirements in the early 1880s, industrial chemists began to organise for their own interest in the Society of Chemical Industry (founded in 1881). The Society was meant to be a learned society dealing with technological topics. Its brief was to discuss what the Chemical Society patently did not. It too was a coalition, and it too was dominated in its early years by an academic approach. Roscoe, apostle of research, was the Society's first president. The Society subscribed to the view that what the chemical industry did, and ought to do more of, was to apply pure science to practical problems.[30]

If the Society could be considered as the learned society of industrial chemistry, the issues of qualification and what was the most important sort of knowledge for the manufacturing chemist were also of concern. There were efforts to establish a companion professional institution. In response to pressure from vocal industrial members of its council, the Institute of Chemistry had attempted to launch an examination in technical chemistry to serve as a qualification for industrialists.[31] However, the examination never took off – it had been essentially a postgraduate qualification and was not attractive to its constituency. In 1913 the Institution of Chemical Technologists was launched. It set out to be the professional body for works chemists and other chemists employed in industry, an area in which its proponents felt the Institute of Chemistry had failed. Neither was this institution a success. However, the importance of chemists during the first World War heightened again the issue of qualification and standards in chemistry. In 1916 H. E. Armstrong, former Professor of Chemistry at the Central Technical College in South Kensington, attempted to promote, from within the Society of Chemical Industry, the body that was eventually established in 1919 as the Federal Council for Pure and Applied Chemistry. This had the specific object of voicing from within the Society the interests of chemical science as distinct from those of the profession of chemistry and the chemical industry.[32] By contrast, in 1922, the Institution of Chemical Engineers was established. A new industrial professional body with a novel intellectual base, its foundation marked a radical departure in the history of the application of chemistry in Britain.[33]

Conclusion

As in modern industry, the relationship of science to technology in mid-Victorian Britain was problematic. Even the application of chemistry, reputedly the most practical science, could not be straightforward. Inevitably innovations in the chemical industry rested on craft ingenuity and engineering skill informed to varying degrees by scientific understanding. The largest and most successful of Britain's chemical industries, the soda industry, entailed the manipulation of bulk quantities of impure solids at high temperatures in unspecifiable conditions. The oft-cited artificial dyestuffs industry in which advances did follow directly from fundamental research in organic chemistry was unusual, not typical. Its economically significant success in Germany at the turn of the century has continued to provide an attractive model of the possibilities of directly applying science that has, however, proved hard to emulate.

Despite the difficulties, there was a genuine desire to apply science to the problems of practice. The possibility had long been believed in and, as exemplary instances multiplied, a growing academic community based its credibility on chemistry's utility. Continued growth depended on convincing still sceptical manufacturers, governments and parents that chemists would be useful advisers and teachers. The chemical curriculum that evolved offered a convincing resolution of the tensions between the potential of science on the one hand and the demands of practice on the other. This curriculum separated the fundamental principles of science, said to be common to the understanding of all chemical endeavours, from the details of individual technologies. Thus whether those technologies were seen to be merely applications of the science previously taught or whether they were understood to have an intellectual structure of their own, academic attention could be focused on what was general – the fundamental principles. The terms 'pure' and 'applied' science came to denote the two sections of the curriculum, the general and the professional. Through a series of intellectually crucial educational debates, the categories developed through the chemical curriculum came to be seen as fundamental to the whole of science. The main institutional outcome of these debates was the creation of a new career: full-time science teaching. For that profession, even the vocational training was defined as pure science.

Regarded as not only useful in a vocational sense, chemistry was widely accepted as part of a liberal education. The process of research in particular, it was argued, developed the mind at the same time as it might discover potentially useful truths. Research training was thus an important part of the core chemical curriculum; ability to teach research came to define the academic's expertise. It was also argued that research methods learnt in academe could be applied in practice. Research too could thus be understood in terms of the categories of pure and applied science which were rooted in the structure of the curriculum. These categories also served to clarify the relationships between various classes of practitioners of chemical skills. Academics, professionals and manufacturers each had their role. Though fewest in number, the academics were increasingly influential. The respect with which they were regarded on account of their research attainment, educational power and institutional resources put them at the top of a new hierarchy of the chemical community. The status of the academics, and of the pure chemistry which they represented, depended in turn on the belief that the application of their science could be seen in practice. The success of the discipline, thus defined, contributed to the acceptance of the terms pure and applied science as fundamental categories. Equally the categories themselves were an intrinsic part of the social and intellectual development of the discipline of chemistry in mid-nineteenth-century Britain.

Appendix A
Biographies

Rather than separately footnoting the biography of each person cited in the text, a collective list with bibliographic notes is given in this appendix. It covers the most significant people mentioned in the text. A source likely to be accessible to the reader is given. That means, in order of precedence, the text of the book itself, the *Dictionary of National Biography*, J. R. Partington, *A History of Chemistry* vol. 4 (London, Macmillan, 1964) (Partington), obituary in the *Journal of the Chemical Society*, any other. Since the first three of these are indexed, no specific page numbers are provided here. Brief identifications are given with each person. Where no more is known than is given in the text, there is no entry.

F. Accum (1769-1835), *DNB*
Pioneer professional chemist
A. Aikin (1773-1854), *DNB*
Secretary, Society of Arts
A. H. Allen (1846-1904), *J. Soc. Chem. Ind.*, **23** (1904), p. 742
Analytical chemist
J. B. Anderson (c. 1785-?), Census 1851, GB, Public Record Office: H.O. 107/1574, f. 169
Scottish and London soap manufacturer
H. E. Armstrong (1848-1937), *DNB*
Professor of Chemistry at Central Technical College
C. Babbage (1792-1871), *DNB*
Mathematician and critic
J. Banks (1743-1820), *DNB*
President of the Royal Society
T. H. Barker (c. 1814-1876), *Yorkshire Gazette*, 18 January 1876
York surgeon and professional chemist
P. P. Bedson (1853-1943), *J. Roy. Inst. Chem.*, 1943, p. 126
Professor of Chemistry, Armstrong College, Newcastle
J. C. Bell (1839-1913), *J. Chem. Soc.*, **105**, pt. 1 (1914), pp. 1193-5
Chemist to Peter Spence
I. L. Bell (1816-1904), *DNB*

Steel and chemical manufacturer
  H. E. Bessemer (1813-1898), *DNB*
Pioneer of steel manufacture ·
  J. Berzelius (1779-1848), see text
Leading Swedish chemist
  G. Birkbeck (1776-1841), *DNB*
Pioneer of Mechanics' Institutes movement
  J. Black (1728-1799), *DNB*
Scottish chemist
  W. T. Brande (1788-1866), *DNB*
Professor of Chemistry at the Royal Institution
  D. Brewster (1781-1868), *DNB*
Scottish Natural Philosopher
  H. Brunner (1838-1916), see text
Merseyside alkali chemist
  H. L. Buff (1828-1872), Poggendorff, *Biographisch-Literarisches Handbuch*
Assistant to Stenhouse, moved to Goettingen
  J. L. Bullock (1814-1906), see text
A founder of the RCC
  F. C. Calvert (1819-1873), *DNB*
Manchester professional chemist
  H. Caro (1834-1911), Partington
German dye chemist
  W. Cavendish, Duke of Devonshire (1808-1891), *DNB*
Chairman of 'Devonshire Commission'
  M. E. Chevreul (1786-1889), Partington
French industrial chemist
  R. C. Clapham (1823-1881), *J. Chem. Soc.*, **41** (1882), p. 236
Tyneside alkali manufacturer
  T. Clark (1801-1867), Partington
Professor of Chemistry, Marischal College, Aberdeen
  H. Cole (1808-1882), *DNB*
Secretary, Science and Art Department
  J. T. Cooper (1790-1854), *Quart. J. Chem. Soc.*, **8** (1856), pp. 109-10
London professional chemist
  H. Croft (1820-1882), J. King, *McCaul; Croft; Forneri* (Toronto, Macmillan,1914)

Professor of Chemistry at Toronto
  W. Crookes (1832-1919), *DNB*
London professional chemist
  W. Crum (1796-1868), *J. Chem. Soc.*, **21** (1868), pp. xvii-xviii
Clydeside calico printer
  W. Cullen (1710-1790), *DNB*
Scottish professor of chemistry
  J. Dale (1815-1889), *J. Chem. Soc.*, **57** (1890), pp. 446-7
Manchester manufacturer of synthetic dyes
  J. G. Dale (1840-1871), *J. Chem. Soc.*, **25** (1872), pp. 344-5
Son and partner of J. Dale
  J. Dalton (1766-1844), see text
Pioneer of atomic theory
  J. F. Daniell (1790-1845), *DNB*
Professor of Chemistry at King's College London
  C. G. B. Daubeny (1795-1877), *DNB*
Professor of Chemistry at Oxford
  G. E. Davis (1850-1906), see text
Chemical engineer
  H. Davy (1778-1829), see text
Leading English chemist
  H. Deacon (1822-1876), Partington
Pioneering soda manufacturer
  H. De la Beche (1796-1855), *DNB*
First Director, Geological Survey
  J. Donnelly (1834-1902), *DNB*
Inspector of Science in the Science and Art Department
  T. Everitt (1803-1846), *Mem. Proc. Chem. Soc.*, **3** (1845-7), p. 141
London professional chemist
  M. Faraday (1791-1867), *DNB*
Professor of Chemistry at the Royal Institution
  W. Forster (1818-1886), *DNB*
Promoter of educational reform
  G. Fownes (1815-1849), *DNB*
Professor of Practical Chemistry at University College London
  E. Frankland (1825-1899), see text
Professor of chemistry at Owens College, St Bartholomew's Hospital
and the Royal College of Chemistry
  J. Gardner (1804-1880), see text

A founder of the RCC
  W. Gossage, (1799-1877), *J. Chem. Soc.*, **33** (1878), p. 229
Pioneer soda manufacturer
  C. Graham (1836-1909), *J. Chem. Soc.*, **97**, pt. 1 (1910), pp. 677-80
Industrial chemist and Professor of Chemical Technology at
University College London
  T. Graham (1805-1869), see text
Professor of Chemistry at University College London
  W. Gregory (1803-1858), *DNB*
Professor of Chemistry at Edinburgh
  J. F. W. Herschel (1792-1871), *DNB*
Leading astronomer
  A. Hofmann (1818-1892), Partington
Professor of Chemistry at the RCC
  L. Howard (1772-1864), *DNB*
Pharmaceutical manufacturer
  F. Hudson (1837-1866), *J. Chem. Soc.*, **20** (1867), p. 382
Manchester consulting chemist
  J. Hudson (c. 1812-1884), *J. Chem. Soc.*, **47** (1885), p. 331
Consulting chemist and educationalist
  T. H. Huxley (1825-1895), *DNB*
Professor of Natural History at the Royal School of Mines and
promoter of technical education
  J. F. W. Johnston (1796-1855), *DNB*
Reader in Chemistry at Durham
  J. Joule (1818-1889), *DNB*
Manchester brewer and natural philosopher
  R. Kane (1807-1890), *DNB*
Director of Museum of Irish Industry, Dublin
  A. Kekulé (1829-1896), Partington
Pioneer of the theory of the structure of organic chemicals. Professor
of Chemistry at Ghent
  S. Kipping (1863-1949), Partington
Professor of Chemistry at University College, Nottingham
  H. Kolbe (1818-1884), Partington
Professor of Chemistry at Marburg
  J. B. Lawes (1814-1900), *DNB*
Superphosphate manufacturer and patron of the agricultural research
laboratory at Rothamsted.
  J. Lockyer (1836-1920), *DNB*

Founder editor of *Nature*
   G. Lowe (1788-1868), *Min. Proc. Inst. Civil Eng.*, **30** (1869-70), pp. 442-5.
Chief Engineer to Gas Light & Coke Co.
   G. Lunge (1839-1923), Partington
Alkali works manager and later Professor of Chemistry at the ETH in Zurich
   F. Lunn (1795-1839), Venn, *Alumni Cantabridgiensis*, pt. II, vol. 4, p. 235
Cambridge mineralogist
   C. Macintosh (1766-1843), *DNB*
Chemical manufacturer
   A. J. F. Marreco (1836-1881), *J. Chem. Soc.*, **41** (1882), p. 238
Professor in Chemistry at the Newcastle College of Science
   J. Mercer (1791-1866), *DNB*
Lancashire calico printer
   W. A. Miller (1817-1870), *DNB*
Professor of Chemistry at King's College London
   R. Montagu, Lord (1825-1902), *DNB*
Vice-President of the Committee of Council
   R. Murchison (1792-1871), *DNB*
Director of the Geological Survey
   J. Muspratt (1793-1886), *DNB*
Pioneer of soda manufacture in Britain
   J. S. Muspratt (1821-1871), *DNB*
Founder of Liverpool College of Chemistry
   J. Napier (1810-1884), *J. Chem. Soc.*, **47** (1885), pp. 333-4
Supervisor for gilding to Elkington
   W. Neild (c. 1789-1864), W E A Axon, *Annals of Manchester* (London, John Haywood, 1886), p. 292
Manchester calico printer
   B. Newlands (1842-1912), *J. Chem. Soc.*, **103**, pt. 1 (1913), pp. 764-5
Chemist to sugar refiners
   W. Odling (1829-1921), *J. Chem. Soc.*, 119, pt. 1 (1921), pp. 553-64
Lecturer in chemistry at St Bartholomew's Hospital
   H. L. Pattinson (1796-1858), see text
Tyneside chemical manufacturer

J. Pattinson (1828-1912), *J. Chem. Soc.*, **103**, pt. 1 (1913), pp. 765-6

Tyneside professional chemist and assistant to H. L. Pattinson
  F. Penny (1816-1869), Partington

Professor of Chemistry at the Andersonian Institution, Glasgow
  W. H. Pepys (1775-1856), *DNB*

Chemical Instrument maker
  J. Percy (1817-1889), *DNB*

Professor of Metallurgy at Royal School of Mines
  W. H. Perkin (1838-1907), *DNB*

Pioneer of coal tar dyes
  R. Phillips (1778-1851), *DNB*

Early London professional chemist
  L. Playfair (1819-1886), see text

Academic and statesman of science
  D. Price (d. 1888), *J. Chem. Soc.* 55 (1889), p. 294

Practical and manufacturing chemist
  W. Prout (1785-1850), *DNB*

Medical chemist
  T. Richardson (1816-1867), *DNB*

Newcastle professional chemist
  E. Riley (1831-1914), *J. Chem. Soc.*, **107**, pt. 1 (1915), pp. 586-8

Iron and steel chemist
  E. Ronalds (1819-1889), *J. Chem. Soc.*, **57** (1890), p. 456

Lecturer in chemistry, London and Galway
  H. E. Roscoe (1800-1836), see text

Professor of Chemistry at Owens College
  J. S. Russell (1808-1882), *DNB*

Engineer and promoter of educational reform
  B. Samuelson (1820-1905), *DNB*

Ironmaster and MP; chairman of the 'Samuelson' Committee and Commission
  J. Sebright (1767-1846), *DNB*

Landowner and amateur chemist
  W. Sharpey (1802-1880), *DNB*

Professor of Physiology at University College London
  J. K. Shuttleworth (1804-1877), *DNB*

Educational reformer
  J. Sinclair (1754-1835), *DNB*

Agriculturist and author of survey of Scotland
  R. A. Smith (1817-1884), *DNB*
First Alkali Inspector
  W. W. Smyth (1817-1890), *DNB*
Director of the Geological Survey
  F. Spence (1837-1907) see text
Son of Peter Spence, alum manufacturer
  P. Spence (1806-1883), *J. Chem. Soc.*, **45** (1884), pp. 622-3
Alum manufacturer
  J. Stenhouse (1809-1881), *DNB*
Lecturer in Chemistry at St Bartholomew's Hospital
  D. Stone (1820-1873), *J. Chem. Soc.*, **27** (1874), pp. 1202-3
Lecturer in chemistry in Manchester
  J. Thomson (1774-1852), *Quart. J. Chem. Soc.*, **4** (1852), pp. 347-8
Clitheroe calico printer
  R. D. Thomson (1810-1864), *DNB*
Lecturer in chemistry in Glasgow and London
  T. Thomson (1773-1852), see text
Professor of Chemistry at Glasgow
  E. Turner (1798-1837), *DNB*
Professor of Chemistry at University College London
  J. Tyndall (1820-1893), *DNB*
Professor of Natural Philosophy at the Royal Institution
  A. Ure (1778-1857), *DNB*
Professor of Chemistry at the Andersonian and consultant
  A. Voelcker (1822-1884), *DNB*
Analytical chemist
  H. Warburton (1784-1858), *DNB*
Reforming member of Parliament
  R. Warington (1807-1867), *DNB*
Founding Secretary of the Chemical Society
  J. T. Way (1821-1884), *J. Chem. Soc.*, **45** (1884), pp. 629-30
Professor of Chemistry at the Royal Agricultural College
  J. Wedgwood (1730-1795), *DNB*
Pottery manufacturer
  W. Weston (1839-1914), *J. Inst. Chem.*, 1915, p. 29
Admiralty chemist
  W. Whewell (1794-1866), *DNB*
Cambridge mathematician
  A. Williamson (1829-1890), see text

Professor of Chemistry at University College London
  J. Wilson (1812-1888), *DNB*
Professor of Chemistry at Edinburgh
  W. H. Wollaston (1766-1828), *DNB*
Leading London chemist
  J. Young (1811-1883), *DNB*
Pioneer of oil shale extraction

Appendix B
Chemical Society analysis

The anonymous editor of the 1896 *Jubilee of the Chemical Society of London* noted that there were undoubted inaccuracies in the membership tally but that the paucity of records made correction impossible. This problem is met repeatedly in the examination of the Society. There are discrepancies between annual lists of elections, notices of elections at meetings and the membership lists. Occasionally members whose elections are noted as void in the manuscript minutes turn up in membership lists. Such problems do not seriously affect overall totals or conclusions; however, they do mean that individual figures of membership are hard to obtain as accurately as one might expect. Given this uncertainty, a small amount of double counting has been allowed in the case of the few who joined in the 1840s, dropped their membership and rejoined in the 1850s.

Classification by the historian introduces its own uncertainties. Analysis of locational distribution is confronted by the uncertain definition of areas – where, for instance, did London stop? There is also the problem of individuals moving. German visitors such as Kekulé joined the Society during their stay in London and then returned to Germany. The problem, is however, not overwhelming if the significance of the figures is borne in mind. Only the broadest impressionistic conclusions can be drawn but they justify presentation of the figures.

Categorisation of men into occupational groups is a particular hazard. The classification of chemists in the nineteenth century was, as is argued in the text, truly ambiguous. A search for the 'correct' classification of each individual would not only be fruitless, it would also be misconceived. Moreover over the first thirty years of the Society's life classifications changed as a cadre of 'academic' chemists emerged. For this reason the prosopography has not integrated the 1840s with the two subsequent decades.

Finally of course many members remain anonymous despite

detailed and ever less fruitful enquiries. Nevertheless, of the 295
members elected from 1841 to 1850, 94% have been classified. Of the
649 members elected to the Chemical Society from 1850 to 1870, 77%
have been identified.

Education is a much more elusive characteristic than occupation.
Information about people was obtained from directories and
obituaries. The former hardly ever provided a clue to education. Since
academics were much more likely to win obituaries than their
colleagues in industry, the training of manufacturers is an especially
uncertain area. It is therefore especially interesting that despite the
vagaries of biographical information, fully 40% of the 238
manufacturers had an identifiable chemical education.

Publication data were obtained from the Royal Society's *Catalogue
of Scientific Papers*. Incomplete as this undoubtedly is, it provides data
which are certainly sufficiently reliable for the purposes here.

Appendix C
Students at the Royal College Chemistry

The Royal College of Chemistry had two distinct phases during the period under discussion. From 1845 to 1853 it was an independent body with an important role in training medical students. Thereafter it joined the Royal School of Mines. Emphasis on casual teaching in both periods is striking. In the period 1845 to 1853 there were 356 students of whom barely 20% (seventy-seven) attended for more than a year and about the same number (seventy) had published a paper by 1883. In the period 1853 to 1870 there were 469 occasional students of whom 20% (ninety-four) stayed more than a year and again just less (eighty-six) published a paper by 1883.

The students who enrolled for just one term in the RCC after 1853 remain very obscure; however, the absence of their names from subsequent Medical Registers indicates that few were on a medical path. However, of the 311 who stayed more than a term, 60% (180) have been identified and of those 50% (ninety-three) entered branches of manufacturing. This is certainly an underestimate of the role of manufacturing careers, because of the paucity of appropriate biographical sources. Nevertheless it does indicate the significance of industry for the alumni of the college and is consistent with the relative lack of of scientific authors among the alumni, though, as is discussed in Appendix D, few became patentees either. The Royal College of Chemistry also enrolled 237 students who had matriculated in the Royal School of Mines. Their subsequent occupations tended to be characteristic of the mining interests of the School, with students becoming mining engineers, mineralogists and geologists, though a few are recorded as having become managers of works.

Appendix D
Patents

The analysis of patent literature used two quite different kinds of sources.

Overall summaries

The overall series of patents was used to construct a time series of increase in number of patents overall. Annual totals for the number of patents before 1852 were taken from Bennet Woodcroft, *Titles of Patents of Inventions Chronologically Arranged. From March 2, 1617 to October 1, 1852* (London, Queen's Printing Office, 1854) and for years after 1852 the year's number of patents in the appropriate volume of the *Chronological List of Patents Applied For and Patents Granted*. The list of Chemical Society patentees in 1860, '61, '70 and '71 was created by scanning the annually produced *Alphabetical Index of Patentees of Invention*.

Class 40

Patents were divided into subject area by Bennet Woodcroft and abstracts of all the patents in each area were published together. Such classes of patents provide a nineteenth-century classification system for the modern historian. It might be anticipated that chemists contributed patents across a wide area. However, most of the classes were so broad that 'chemical' patents constituted but a small part. Class 40 was chosen because it was the only class of patents which had a significant density of Chemical Society members as patentees and therefore could be considered as 'chemical' in more than name. The listing of patents on which the analysis is based was published in two series: (i.) *Abridgments of Specifications Relating to Acids, Alkalies, Oxides and Salts A.D. 1622-1866* (London: Office of the Commissioners of Patents for Inventions, 1869), ii. The second part of the above series published in 1888 was divided into three separate volumes each encompassing a specific chemical area, though a single patent might be abstracted in more than one volume. The parts are: Division I, Acids, Chlorine, Sulphur &c; Division II, Alkalies, Oxides and Salts; Division III, Benzene Derivatives and other Carbon Compounds.

Analysis

1. Patents. Since the single series ends in 1866, it would be unprofitably laborious to continue the count of patents beyond that date.

2. Patentees. Patentees have been identified from the biographical information provided in the *Chronological List of Patents*, which was taken originally from the patents themselves. This series terminates in 1875 and that date determined the end date of the analysis. Most of the analyses are based only on the period 1853 to 1875 because the 1852 Patent Act introduced a major shift in patent numbers. Moreover only after the 1852 Act did foreign patentees become generally identifiable.

The analysis of Class 40 patentees was intended to cover all those who patented in the area before the close of 1875. However, there are a series of exceptions.

a. People who had published patents in the area before 1830. Since each patentee was associated with the date of entry into the index, pre-1830 patentees have been excluded. In any case their number was very small.

b. People who were publishing patents that were communications from others. Before 1852 communications were described as such but originators were not on the whole mentioned. After the 1852 Patent Act originators were generally given. For communications, then, only the originators of the innovation rather than the actual patentees have been included in the list.

c. People whose patents were described as 'void' or as denied 'provisional protection' in the *Abridgments of Specifications* have been, as a convention, omitted. Their numbers are in any case small. All other patents have been retained whatever qualifications are noted.

Patentees defined thus have been divided at two levels. First they have been divided into 'British' and 'Foreign'. The foreign patentees of various countries have been summed to give an indication of the significance of various national contributions but have not otherwise been analysed. British patentees have been classified by occupation. The *Chronological List* allowed the identification of about 53% of the

British patentees. We were particularly interested in the categories of 'chemist' and 'engineer'. Two definitions of chemist were used. The less restrictive included those calling themselves 'chemist' with some qualifier as in 'manufacturing chemist', 'analytical chemist' and 'consulting chemist', and the scarcely separable classes of 'chemical manufacturer', 'alkali manufacturers' (of whom there were two) and similar numbers of 'colour manufacturers' and 'alum manufacturers'. Included also were those few who styled themselves FCS (Fellow of the Chemical Society) or gave the Royal College of Chemistry as the address. Using this definition there were 250 'chemists' in the period 1853 to 1875. The more restrictive definition of chemists excluded the manufacturers and reduced the number to 121. As will be seen, in fact while there were differences in growth rates between the two categories of chemists they are not sufficient to affect the conclusions drawn. Over the period 1853 to 1875, 103 self-styled 'engineers' have been identified.

Statistical analysis

As could only be appropriate for such a crudely constructed sample the analysis conducted has been extremely simple. No inference about underlying patterns has been drawn. The only object has been to find a simple way of describing trends in the number of new patentees in different classes. In order to identify major differences in growth patterns we have used the simple device of finding the best-fit exponential curves whose formulae can then be compared. We are therefore not claiming that there was an underlying mechanism by which the number of patentees was actually growing exponentially. The implicit assumption is rather that reduction to exponential curves is not grossly misleading.

The basic equation for compound growth rate is $P = KB^T$, where P represents the magnitude of the growing population, T is time, K is a characteristic constant, and B is a characteristic constant whose value is related to the more commonly cited percentage growth rate R, by the formula $R = 100(B - 1)$. The exponential equation can be reduced to linear form by taking logarithms, yielding $\ln P = \ln K + (\ln B)T$. Thus if values of $\ln P$ are plotted against T a straight line would follow. The formula of this, and hence the growth rate, could be easily determined by the method of least squares. If the values do not all fall on the line, a correlation coefficient can be calculated. The square of this coefficient gives the proportion of the variance between the values (of the logarithm of P) accounted for by the calculated line. It is therefore an indication of the accuracy of the curve's fit.

The above procedure was carried out for various sets of Class 40 patentees. Below are given the equation of log number of patentees against time for each set of patentees, the correlation coefficient for the line, and percentage growth rate entailed by the line's formula. Since the curves are so crude, growth rates are rounded in the text to the nearest whole number percentage.

For British patentees 1853-1875
$\ln P = 2.979 + O.0527 \times$ Years
r (correlation coefficient) = 0.79
Growth rate = 5.4% per annum

For foreign patentees 1853-1875
$\ln P = 2.406 + 0.0633 \times$ Years
$r = 0.92$
Growth rate = 6.5% per annum

For British engineers 1853-1875
$\ln P = -0.008 + 0.1099 \times$ Years
$r = 0.88$
Growth rate = 11.6% per annum

For British chemists (loose definition) 1853-1875
$\ln P = 1.988 + 0.0286 \times$ Years
$r = 0.45$
Growth rate = 2.9% per annum

For British chemists (tight definition) 1853-1875
$\ln P = 1.108 + 0.0383 \times$ Years
$r = 0.39$
Growth rate = 3.9% per annum

Number of new patentees Class 40
1830-1875

Total	British	Foreign	French	German	US
1704	1122	582	316	90	123

Chemical Society and RCC patentees.

The samples of Chemical Society members and RCC alumni were compared with the Class 40 set of patentees.

The Chemical Society membership could be compared in two ways: the proportion of the Chemical Society membership who patented and the proportion of patentees who joined the Chemical Society. For the former analysis Chemical Society membership to 1870 was compared to Class 40 patenting to 1883. For this purpose, Part 3 of the Class 40 index was consulted. It was found that out of the 944 Chemical Society members to 1870, 139 had a patent in Class 40 by 1883. That is, approximately one in seven of the Chemical Society members had patented. To examine the Chemical Society membership of patentees, the patentees to 1866 were compared to Chemical Society membership up to 1870. Out of 623 British patentees in Class 40 to 1866, ninety-three had joined the Society. Of these sixty-five described themselves as 'chemists', thirteen, gave no description, six were 'FCS', 'alum manufacturer', 'dye manufacturer', 'soap manufacturer' and 'PhD', four were 'engineers' and four doctors. In the period 1853-1866, 127 patentees in Class 40 had called themselves 'chemist'. Therefore about half of them had joined the Society.

In conclusion the membership of the Chemical Society and the number of patentees in the central chemical area were of the same order of size, but overlapped only to a very limited extent. At the same time those with a specially chemical identity among the patentees did seem to have joined the Society.

Patenting by alumni of the Royal College of Chemistry in Class 40 was rather low. Of the occasional students only twenty-four had patented in the class by 1883. An analysis of the wider patenting of alumni was attempted, but there are considerable problems of distinguishing patentees with the same name. So the results could only be very tentative. Approximately thirty alumni, at a conservative estimate, seem to have patented. From this one might deduce that of the order of 10%, and probably less, of the RCC students later patented.

It may seem surprising that alumni of a school which included such famous innovators as W. H. Perkin should have been so sparsely represented among patentees. However, the subject with which they are particularly associated, synthetic dyestuffs, was in the 1860s still a relatively minor part of technology and in fact, after initial discoveries, was not dynamic till the late 1860s. In addition to the list of patents of chemicals in Class 40, the specialised category of dyestuffs patents was also examined. This could not be done in quite the same way because Class 14, 'Bleaching, Dyeing, and Printing Calico and other fabrics and yarns' covers an enormous span, largely beyond the chemistry of dyestuffs. So instead the more detailed retrospective subject index for 1861-1910 was surveyed for the years 1861 to 1875 (*Patents for Inventions. fifty Years Subject Index 1861-1910*, London, HMSO, 1915). Two categories within that index were chosen as of special interest. Class 2 iii is entitled 'Dyes and Hydrocarbons' and Class 15 i 'Dyeing equipment' and ii 'Dyeing Processes'. Again for each of these two sets first-time patentees within the group were identified and dated. Because the index only begins in 1861 it would of course be impossible to use it to exclude patentees already established in the area before 1861. However, if they were not excluded one would tend to overestimate the number of first-time patentees in the early 1860s. So the list of patentees arrived at from this list was compared with the Class 14 lists, Parts 1 and 2 covering the period before 1861, and any name that appeared in Class 14 was eliminated from the set. The total number of new patentees within the group remaining was as follows:

Year	Corrected figure	Uncorrected figure
1861	10	10
1862	9	10
1863	8	14
1864	8	10
1865	6	8
1866	8	8
1867	7	8
1868	3	4
1869	6	6
1870	4	6
1871	4	4
1872	8	9
1873	5	5
1874	10	10
1875	7	8

The values for classes 15 i and ii turned out to be too small to analyse. Only half as many patents were registered in this class between 1861 and 1875 as in Class 2 iii (212 compared to 344).

This pattern is similar to that detected by A. G. Bloxam in his 'Patent Law in Relation to the Dyeing Industry', in Walter Gardner, ed., *The British Coal Tar Industry: Its Origins, Development and Decline* (London, Williams and Morgate, 1915), pp. 265-79. Our work suggests that when corrected for established inventors, even the small peak of 1863 disappears and there seems to have been a continuous decline of new inventors in the area during the 1860s.

Notes

Abbreviations: Where appropriate, certain journal titles have been abbreviated in accordance with the guidance given in *Handbook for Chemical Society Authors*, Chemical Society Special Publications 14 (London, Chemical Society, 1960). In addition, we have used the standard abbreviations *DNB* for *Dictionary of National Biography* and *P.P.* for *Parliamentary Papers*.

Introduction

1. 'Education for capability', *The Times*, 27 February 1980. One recent analysis of the roots of this anxiety is M.J. Wiener, *English culture and the decline of the industrial spirit, 1850–1980* (London, Hutchinson, 1981). Our approach is rather different.

2. C.W. Sherwin and R.S. Isenson, *First interim report on Project Hindsight: Summary*. Report AD 642–200 (Washington D.C., Office of Director of Defense Research and Engineering, 1966).

3. IIT Research Institute, *Technology in retrospect and critical events in science (TRACES)*, NSF Contract NSF–C535 (2 vols., Chicago, IIT Research Institute, 1968). For a discussion of 'Hindsight' and 'Traces', see K. Kreilkamp, 'Hindsight and the real world of science policy', *Sci. Studs.*, 1 (1971), 43–66.

4. Great Britain. Parliament. *A framework for government research and development*, Cmnd 4814 (London, HMSO, 1971).

5. For an exposition of the philosophical assumptions underlying the Dainton report, see '"The future of the Research Council System"', Report of a C.S.P. Working Group under the Chairmanship of Sir Frederick Dainton' [to give the 'Dainton Report' its full name] in *Framework for government research*, para 8.

6. Lord Rothschild, 'The organisation and management of Government R. & D.' [the Rothschild Report] in *Framework for government research*, para 25.

7. For the debate after the publication of the Dainton and Rothschild reports, see Philip Gummett, *Scientists in Whitehall* (Manchester, Manchester University Press, 1980), pp. 195–213.

8. For a summary of the literature, see E. Layton, 'Conditions of technological development' in Ina Spiegel-Rösing and Derek De Solla Price eds., *Science, technology and society: A cross-disciplinary perspective*, International Council for Science Policy Studies (London, Sage Publications, 1977), pp. 197–222.

9. E.T. Layton, 'Mirror-image twins: The communities of science and technology in nineteenth-century America', *Technology and Culture*, 12 (1971), 562-80.

10. We have purposely kept short our references to countries other than Britain. Such incidental discussions of highly complex national situations as might fit within our scope could hardly be other than superficial and potentially misleading. For entry into the labyrinthine literature on German higher education in the nineteenth century, see Charles E. McClelland, *State, society, and university in Germany, 1700-1914* (Cambridge, Cambridge University Press, 1980); and Fritz K. Ringer, *The decline of the German mandarins: The German academic community, 1890-1933* (Cambridge, Mass., Harvard University Press, 1969). On both Germany and France, see *idem, Education and society in modern Europe* (Bloomington, Ind., University of Indiana Press, 1979).

11. For studies of the development of the French scientific system that parallel this study of British circumstances, see Robert Fox and George Weisz, eds., *The organization of science and technology in France, 1808-1914* (Cambridge, Cambridge University Press, 1980) and John Hubbel Weiss, *The making of technological man: The social origins of French engineering education* (Cambridge, Mass., MIT Press, 1982).

12. For access to the literature on the American educational system and the pattern of scientific development in the nineteenth century, see Alexandra Oleson and John Voss, eds, *The organization of knowledge in modern America, 1860-1920* (Baltimore, Md., The Johns Hopkins University Press, 1979). One particular monograph worth pointing out here because of the parallels to the British situation described below is Monte A. Calvert, *The mechanical engineer in America, 1830-1910: Professional cultures in conflict* (Baltimore, Md., The Johns Hopkins University Press, 1967). See also, George H. Daniels, 'The pure-science ideal and democratic culture' *Science*, **156** (1967), 1699-1705.

13. D.S.L. Cardwell, *The organisation of science in England*, 2nd ed. (London, Heinemann Educational Books, 1972), pp. 111-55. Cardwell's work has emerged as the classic among a surprising number of studies of the nineteenth-century scientific system carried out during the 1950s. It has still not been surpassed. For comprehensive lists of the numerous theses and books on the educational aspects, see W.H. Brock, 'From Liebig to Nuffield: A bibliography of the history of science education, 1839-1974', *Studs. in Sci. Ed.*, **2** (1975), 67-99; Christine M. Heward and John Naylor, 'Industry, cleanliness and Godliness: Sources for and problems in the history of scientific and technical education for the working classes, 1850-1910', *ibid.*, **7** (1980), 87-128; and Roy MacLeod and Russell Moseley, 'Breadth, depth, and excellence: Sources and problems in the history of university science education in England, 1850-1914', *ibid.*, **5** (1978), 85-106. For the importance of 1870 as a turning point in the development of British science, see Margaret Gowing, 'Science, technology, and education: England in 1870', *Notes Recs. Roy. Soc.*, **32**, pt. i (1977), 71-90.

14. See Fritz K. Ringer, 'The German academic community' in *The organization of knowledge in modern America*, pp. 409-29.

15. An important recent exposition of the complexity of early nineteenth-century British science that challenges traditional generalisations and over-simplifications is Ian Inkster and Jack Morrell, eds., *Metropolis and province: Science in British culture, 1780-1850* (London, Hutchinson, 1983).

16. William Whewell coined the word 'scientist' to describe the variety of participants in the British Association meetings, see his 'Mrs. Somerville on the connexion of the sciences', *Quart. Rev.*, **51** (1834), 54-68. For the history of the word, see Sidney Ross, '*Scientist:* The Story of a Word', *Ann. Sci.*, **18** (1962), 65-85. The context of the neologism is given in Jack Morrell and Arnold Thackray, *Gentlemen of science: Early years of the British Association for the Advancement of Science* (Oxford, Clarendon Press, 1981), p. 20.

17. The distinction between disciplines, specialties and research areas has been most carefully formulated by Richard Whitley in 'Cognitive and social institutionalization of scientific specialties and research areas', in *idem*, ed., *Social processes of scientific development* (London, Routledge and Kegan Paul, 1974), 69-95; and *idem*, 'Umbrella and polytheistic disciplines and their elites', *Soc. Studs. Sci.*, **6** (1976), 471-97. See also Daryl Chubin, 'The conceptualization of scientific specialties', *Sociol. Quart.*, **17** (1976), 448-76.

18. Here we are seeing chemistry as what Whitley has called an 'umbrella discipline', Richard Whitley, 'Umbrella and polytheistic disciplines'.

19. Karl Hufbauer, *The formation of the German chemical community* (Berkeley, Calif., University of California Press, 1982).

20. Robert E. Kohler, *From medical chemistry to biochemistry: The making of a biomedical discipline* (Cambridge, Cambridge University Press, 1982).

21. Daniel J. Kevles, *The Physicists* (New York, Knopf, 1977).

22. 'Sixteenth Report of the Science and Art Department of the Committee of Council on Education', *P.P.* 1868-69 [4136] XXIII. 131, App. B, p. 63.

23. The sorts of communities we have been dealing with numbered roughly a thousand by the 1870s. For growth to the turn of the century see Wyndham Hulme, *Statistical bibliography in relation to the growth of modern civilization: Two lectures delivered in the University of Cambridge in May 1922* (London, pub. by the author, 1923), Tables 1, 2, and 3.

24. Owen Hannaway, *The chemists and the word: The didactic origins of chemistry* (Baltimore, Md., The Johns Hopkins University Press, 1975). See also C. Meinel, '"De praestantia et utilitate Chemiae": Selbstdarstellung einer jungen Disziplin im Spiegel ihres programmatischen Schrifttums', *Sudhoffs Archiv*, **65** (1981), 366-89.

25. Lyon Playfair to Henry Cole, 17 March 1853, Cole Papers, Correspondence, Box 15, Victoria and Albert Museum.

Chapter one

1. Francis Bacon, *The New Organon*, ed. F.H. Anderson (Indianapolis, Bobbs, Merrill, 1968), first pub. 1610, Aphorisms, Book 1, para. 81. See Paolo Rossi, *Francis Bacon: From magic to science*, trans. S. Rabinovitch (London, Routledge and Kegan Paul, 1968).

2. See particularly Archibald and Nan L. Clow, *The Chemical Revolution: A contribution to social technology* (London, Batchworth press, 1952); and A.E.

Musson and Eric Robinson, *Science and technology in the Industrial Revolution* (Manchester, Manchester University Press, 1969).

3. Humphry Davy, 'A discourse introductory to a course of lectures on chemistry' (1802) in *Revolutions, 1775-1830,* ed. M. Williams (Harmondsworth, Penguin Books Ltd., 1971), pp. 397-409. The significance of this lecture is discussed in Morris Berman, *Social change and scientific organization: The Royal Institution, 1799-1844* (London, Heinemann Educational Books, 1978), p. 67.

4. On Oxford, see Harold Hartley, 'Schools of chemistry in Great Britain and Ireland – XVI: The University of Oxford', *J. Roy. Inst. Chem.,* **79** (1955), 116-27, 176-84. On Cambridge, see W.H. Mills, 'Schools of chemistry in Great Britain and Ireland – VI: The University of Cambridge', *ibid.,* **77** (1953), 423-31, 467-73.

5. H.H. Bellot, *University College London, 1826-1926* (London, University of London Press, 1929), pp. 14-59.

6. On the London elite, see P.J. Weindling, 'Geological controversy and its historiography: The prehistory of the Geological Society of London', in *Images of the Earth: Essays in the history of the environmental sciences,* eds. L.S. Jordanova and R.S. Porter, BSHS Monographs 1 (Chalfont St. Giles, British Society for the History of Science, 1979), pp. 248-71. The rise of the manufacturers is the subject of A.E. Musson and E. Robinson, *Science and technology.*

7. W.A. Campbell, 'The analytical chemist in nineteenth-century English social history', M.Litt. Thesis, University of Durham, 1971.

8. Arnold Thackray, 'Natural knowledge in cultural context: The Manchester model', *Am. Hist. Rev.,* **79** (1974), 672-709. See also Ian Inkster, 'Culture, institutions and urbanity: The itinerant science lecturer in Sheffield, 1790-1850' in S. Pollard and C. Holmes, *Essays in the economic and social history of South Yorkshire* (Sheffield, South Yorkshire County Council, 1976).

9. Arthur Donovan, 'British chemistry and the concept of science in the eighteenth century', *Albion,* **7** (1975), 131-44. See also J.R.R. Christie, 'The origins and development of the Scottish scientific community, 1680-1760', *Hist. Sci.,* **12** (1974), 122-41.

10. F.R. Cowell, *The Athenaeum: Club and social life in London, 1824-1974* (London, Heinemann, 1975), pp. 10-12.

11. On the spread of literary and philosophical societies, see A. Thackray, 'Natural knowledge', p. 674. On London societies see J.B. Morrell, 'London institutions and Lyell's career, 1820-41', *Brit. J. Hist. Sci.,* **9** (1976), 132-46.

12. M. Foucault, *The order of things* (New York, Vintage Books, 1973), pp. 125-65. See also 'Classification and systematization in the sciences' in E.G. Forbes, ed., *Human implications of scientific advance* (Edinburgh, Edinburgh University Press, 1978), pp. 495-538.

13. P.J. Weindling, 'Geological controversy'.

14. Sir John Sinclair to Arthur Young, 27 September 1811, Add. MSS 35131, f. 192, British Library, London.

15. Harold R. Fletcher, *The Story of the Royal Horticultural Society, 1804-1968* (London, Oxford University Press, 1969).

16. On the founding of the Geological Society and its early years see Horace B. Woodward, *The History of the Geological Society of London* (London, Geological Society, 1907); Martin Rudwick, 'The foundation of the Geological Society of London: Its scheme for cooperative research and its struggle for independence', *Brit. J. Hist. Sci.*, 1 (1962-3), 325-55; Roy Porter, 'The Industrial Revolution and the rise of the science of geology', in *Changing perspectives in the history of science*, eds. Mikulás Teich and Roy Porter (London, Heinemann, 1973), pp. 320-43; and Weindling, 'Geological Controversy'.

17. Rev. George Bellas Greenough, 'Address', *Proc. Geol. Soc.*, 2 (1834-38), p. 42.

18. Rachel Laudan, 'Ideas and organizations in British geology: A case study in institutional history', *Isis*, 68 (1977), 527-38.

19. See for example J.L.E. Dreyer, ed., *History of the Royal Astronomical Society, 1820-1920* (London, Royal Astronomical Society, 1923), p. 6; David E. Allen, *The naturalist in Britain* (London, Allen Lane, 1976); and P.J. Weindling, 'The British Mineralogical Society: A case study in science and social movement' in Ian Inkster and J.B. Morrell, eds., *Metropolis and province: Science in British culture, 1780-1850* (London, Hutchinson, 1983), pp. 120-50.

20. C.R. Weld, *A history of the Royal Society: With memoirs of the Presidents compiled from authentic documents*, (2 vols. London, John W. Baker, 1848), vol. 2, pp. 149-50; see also the obituary of Richard Bright in *Phil. Mag.*, 3rd. ser., 20 (1842), 523-6.

21. Information on the club and its membership is scattered. Davy mentioned it as an ongoing organisation in 1807, H. Davy to W.H. Pepys, 13 November 1807, in John Ayrton Paris, *The life of Humphry Davy* (London, Henry Colburn and Richard Bentley, 1831), p. 179. The latest date found by L.F. Gilbert during research for a never-completed paper on the Club was 1826. He was able to give that date for a letter from H. Davy to J. Bostock ; Gilbert MSS, Box 4, file 1, Enc. A, University College London. The Gilbert MSS include transcripts of many letters as well as Gilbert's notes. Berzelius' visit is recorded in H.G. Söderbaum, ed., *Jac. Berzelius Reseantechningar* (Stockholm, P.A. Norstedt & Söner, 1903), p. 38, entry for 28 July 1812; he reported a membership of fifteen. Dalton's visit is noted in H.E. Roscoe, *John Dalton and the rise of modern chemistry* (London, Cassell & Co., 1895), pp. 169-70. The exclusion of Pepys is noted in Paris, *Life of Humphry Davy*, p. 179; the exclusion of Prout is complained of in Söderbaum, *Jac. Berzelius Reseantechningar*, p. 114. On Warburton's membership, see *ibid.*, p. 38; on Sebright, see W. Babington to A. Marcet, 7 September 1821, Gilbert MSS, Box 4, file 1, Enc. A 13.

22. Chemical topics are noted in A. Marcet to J. Berzelius, 31 October 1812, in H.G. Söderbaum, *Jacob Berzelius Bref* (6 vols., Stockholm, Almquist & Wiksell, 1912-32), vol. 1, pt. III, No. 2, pp. 7-10. For other topics of discussion see Marcet to Leonard Horner, 15 April 1816, in M. Lyell, ed., *Memoir of Leonard Horner, FRS, FGS: Consisting of letters to his family and from some of his friends*, (2 vols., London, Woman's Printing Society, 1890), vol. 1, pp. 92-3.

23. 'Premiums for discoveries and improvements in chemistry, dyeing and mineralogy', *Trans. Soc. Arts*, **37** (1820), pp. xiii-xix.

24. Sir Henry Trueman Wood, *A history of the Royal Society of Arts* (London, John Murray, 1913); see also Derek Hudson and Kenneth W. Luckhurst, *The Royal Society of Arts, 1754-1954* (London, John Murray, 1954), esp. pp. 119-25 and 176-7.

25. Morris Berman, *The Royal Institution*.

26. J.Z. Fullmer, 'Davy's sketches of his contemporaries', *Chymia*, **12** (1978), 127-50.

27. J. Banks to R. Chevenix, 22 March 1805; Dawson Turner Collection, 15, ff. 338-40, British Museum (Natural History), London. For the full story, see M.C. Usselman, 'The Wollaston/Chevenix controversy over the elemental nature of Palladium: A curious episode in the history of chemistry', *Ann. Sci.*, **35** (1978), 551-79.

28. D.F.S. Scott, ed., *Luke Howard (1772-1864): His correspondence with Goethe and his continental journey of 1818* (York, William Sessions Ltd., 1976), p. 4.

29. 'London Chemical Society', *Phil. Mag.*, **25** (1806), 83-6; 'Prospectus of an establishment to be called the London Chemical Society', *Journal of Natural Philosophy, Chemistry and the Arts*, **14** (1806), 268-70.

30. W.H. Brock, 'The Chemical Society, 1824', *Ambix*, **14** (1967), 133-9; C.A. Russell, N.G. Coley, and G.K. Roberts, *Chemists by profession: The origins and rise of the Royal Institute of Chemistry* (Milton Keynes, The Open University Press, 1977), pp. 55-60. For *The Chemist*, see Florence Wall, '"The Chemist" – earliest chemical periodical in English', *Chemist*, **18** (1941), 248-59.

31. H. Davy to J. Children, n.d., Add. MSS 38265, f. 55, British Library.

32. 'Apology and preface', *Chemist*, **1** (1824), vii.

33. See the pioneering work of Everett Mendelsohn, 'The emergence of science as a profession in nineteenth-century Europe', in K. Hill ed., *The management of scientists* (Boston, Beacon Press, 1964).

34. See T.J. Johnson, *Professionals and power* (London, Macmillan, 1972).

35. *Med. Physical J.*, **4** (1800), 280-3; and *ibid.*, **34**, (1815), 259-60. On the role of chemistry in medical education, see J.K. Crellin, 'Chemistry and 18th-century medical education', *Clio Med.*, **9** (1974), 9-21.

36. Apothecaries Hall, *Regulations for the examination of Apothecaries, 1815, ibid.*, 1828; *Regulations to be observed by students intending to qualify themselves to practise as Apothecaries in England and Wales MDCCCXXXV* (London, 1835). Courses are listed in *Medical calendar or student's guide to the medical schools in Edinburgh, London, Dublin, Paris, Oxford, Cambridge, Glasgow, Aberdeen and St. Andrews* (Edinburgh, Maclachlan & Stewart, 1828).

37. Society of Apothecaries, 'Minutes of the Court of Assistants', **12** (1834-45), p. 120, MS 8200, Guildhall, London.

38. Enrolments at the RI are given in 'Report from the Select Committee on Medical Education with the Minutes of Evidence and Appendix', *P.P.* 1834 (602-III) XIII.III. 1, p. 111. Fees are noted in 'Account of the Hospitals

and Schools of Medicine in London for the season of 1840-41, Commencing 1 October 1840', *Lancet* (1840-1), pt. 1, 4-18. Earnings at Guy's are mentioned in J. Bostock to A. Marcet, 23 July 1820, Bostock-Marcet Correspondence, Duke University Medical Center Library, Durham, N.C.

39. Society of Apothecaries, 'Minutes of the Court of Examiners', **4** (1828-1833), November 1830, MS 8239, Guildhall.

40. Royal Institution, Archives, vol. 9, p. 162. On lecturing see I. Inkster, 'Culture, institutions and urbanity'. See also J.N. Hays, 'The London lecturing empire, 1800-50' in I. Inkster and J.B. Morrell, *Metropolis and province*, pp. 91-119.

41. See 'The intellectual resources of London', *Inventors' Advocate, Patentees' Recorder, and Weekly British and Foreign Miscellany of Inventions, Discoveries and the Fine Arts*, **30** (November 1839), 247; and Thomas Coates, *Report of the state of Literary, Scientific and Mechanical Institutions in England* (London, Society for the Diffusion of Useful Knowledge, 1841), Appendix IV, pp. 106-12.

42. J.T. Cooper to the Council of the University of London, 19 October 1827; Applications, College Correspondence, College Archives, University College London.

43. M. Faraday to J. Pelly, 3 February 1836, S.C. MSS 2, Box 4, Institution of Electrical Engineers. On Faraday's consulting, see also *DNB*, and Sylvanus P. Thomson, *Michael Faraday: His life and work* (London, Cassell, 1901), p. 245.

44. Census 1841, Great Britain, Public Record Office: H.O. 107/1083. 5. 9, f. 28. London.

45. Robert H. Kargon, *Science in Victorian Manchester: Enterprise and expertise* (Manchester, Manchester University Press, 1977).

46. [Benjamin Brodie], *Autobiography of the late Sir Benjamin Brodie* (London, Longman, 1865), pp. 90-1. W.T. Brande to M. Faraday, Christmas 1839, S. C. MSS 2, Box 2, Institution of Electrical Engineers.

47. Berman, *Royal Institution*, esp. pp. 100-55.

48. William Murdoch, 'An account of the application of the gas from coal to economical purposes', *Phil. Trans.* , **98** (1808), 124-32.

49. 'A Statement of the Total Number of Retorts, Contents of Gasometers, Coals Consumed, Gas Generated, Coke Produced, and Lamps Lighted, throughout the Metropolis during the Year 1822, by the Different Companies under the Control of the Secretary of State', in 'Gas Light Establishments: Reports of the Gas Light Establishments in the Metropolis', *P.P.* 1823 (193) V. 303, Enclosure 8, pp. 38-9. On the Chartered Gas Company see Stirling Everard, *History of the Gas Light and Coke Company, 1812-1949* (London, Ernest Benn, 1949).

50. Andrew Ure, *Dictionary of arts, manufactures and mines (London, Longman, Brown, 1839)*, p.589.

51. Contemporary conceptions of the technical problems are made clear in Samuel Clegg Jr., *A practical treatise on the manufacture and distribution of coal gas* (London, John Weale, 1841). See p. 46 for a discussion of determining the quality of gas. See also Dean Chandler, *Outline of history of lighting by gas* (n.

pl., n. pub., n.d [c. 1936]). See also W. Matthews, *An historical sketch of the origins, progress, & present state of gas lighting* (London, Rowland Hunter, 1827), pp. 68-71.

52. W.T. Brande, 'Observations on the application of coal gas to the purposes of illumination', *Quart. J. Sci.*, **1** (1816), 71-9.

53. Chartered Gas Company, 'Minutes of the Committee of Works', 10 April 1837 and 7 May 1838, B/GLCC/96, pp. 2 and 151; GLC Record Office, London.

54. David Pollock to the President and Council of the London University, 31 October 1827, Applications, College Correspondence, College Archives, University College London. On the oil gas issue generally see Berman, *Royal Institution*, pp. 145-51, and Everard, *History of the Gas Light and Coke Company*, p. 97. The scientific testimony is recorded in *Minutes of evidence taken before a Committee of the House of Commons on the London and Westminster oil-gas bill, sessions 1824* (London, n.p, n.d). The only copy of this report located so far is in the Goldsmith's Library, Senate House, London, Acc. No. [GL] I3. 823 fol.

55. 'Report from the Select Committee on the Law relative to Patents for Inventions', *P.P.* 1829 (332) III. 415, testimony of John Farey, p. 149.

56. Cited in Andrew Ure, *The Revenue in jeopardy from spurious chemistry* (London, James Ridgway, 1843), p. v. An often cited example of conflicting chemical experts during the 1820s is the case of Severn and King against various insurance companies, see J.Z. Fullmer, 'Technology, chemistry, and the law in early nineteenth-century England', *Technology and Culture*, **21** (1980), 1-28. See also W. Spence, *Patentable inventions and scientific evidence* (London, Stevens and Norton, 1851).

57. On Aikin see *Min. Proc. Inst. Civil Eng.*, **14** (1855), 120-3. On the skills needed to draft patents see William Carpmael, *Law reports of patent cases* (3 vols., London, A. Macintosh, 1843-52), 'preface', vol. 1, pp. iii-vii. See also J. Harrison, 'Some patent practitioners associated with the Society of Arts, 1790-1840', *J. Roy. Soc. Arts*, **30** (1982), 494-8.

58. 'The Chemical Club or Bacchus versus Hermes', in 'Faraday Common-Place Book', vol. 1, n.d., p. 441, S.C. MSS 2, Box 1, Institution of Electrical Engineers.

59. *Yorkshire Gazette*, 18 January 1876.

60. M. Byrne, 'Thomas Richardson, his contribution to chemical education', *Durham Research Rev.*, **7** (1974), 444-7.

61. This figure, which includes books on subjects such as agricultural chemistry, and brewing, is derived from *The classified index to the London Catalogue of Books Published in Great Britain, 1816 to 1851* (London, Thomas Hodgson, 1853). In the seven five-year periods between 1816 and 1851, the number of books in this category were: 10, 15, 20, 16, 19, 32, 35, 25.

62. E.A. Parnell, *The life and labours of John Mercer, the self-taught chemical philosopher: Including numerous recipes used at the Oakenshaw Calico-Print Works* (London, Longman, Green, & Co., 1886).

63. Ian Inkster, 'Science and Mechanics' Institutes, 1820-1850: The case of Sheffield', *Ann. Sci.*, **32** (1975), 451-74 (468); and *idem*, 'The development of a scientific community in Sheffield, 1790-1850: A network of people and interests', *Trans. Hunter's Archaeol. Soc.*, **10** (1973), 99-131.

64. R. A. Smith, 'A centenary of science in Manchester', *Manchester Lit. & Phil. Soc. Mem. Proc.*, ser. 3, **9** (1883), p. 348.

65. Arnold Thackray, 'Natural knowledge', p. 686.

66. Robert Kargon, *Science in Victorian Manchester*, p. 43

67. *Ibid.*, pp. 16-19; R.F. Bud, 'The Royal Manchester Institution' in D.S.L. Cardwell, ed., *Artisan to graduate: Essays to commemorate the foundation in 1824 of the Manchester Mechanics' Institution, now in 1974 the University of Manchester Institute of Science and Technology* (Manchester, Manchester University Press, 1974), pp. 119-33.

68. For this view of Dalton, see Arnold Thackray, *John Dalton: Critical assessments of his life and science* (Cambridge, Mass., Harvard University Press, 1972).

69. D.W.F. Hardie and J. Davidson Pratt, *A history of the modern British chemical industry* (Oxford, Pergamon Press for the Society of Chemical Industry, 1966), p. 15; and Edmund Potter, *'Calico printing as an art manufacture': A lecture read before the Society of Arts, 22 April 1852* (London, John Chapman, 1852), p. 27.

70. For the life of Pattinson, see 'Hugh Lee Pattinson' in H. Lonsdale, *Worthies of Cumberland* (6 vols. London, George Routledge and Sons, 1873), vol. 6, pp. 273-320.

71. H.L. Pattinson to J.B. Anderson, 11 December 1829. Pattinson Letter Book, vol. 1, Wellcome MS 3805, Library of the Wellcome Institute for the History of Medicine, London.

72 H.L. Pattinson to J.B. Anderson, 8 February 1830, *ibid*.

73. For Pattinson's studies of the French language and French science see for example H.L. Pattinson, 'Translation of Papers on the Nature of Fat in the *Annales de Chimie*', Common Place Book No. 1 (1821-4), pp. 51-174, Wellcome MS 1589. See also 'Translation of Several memoirs on the Subject of Saponification', Common Place Book No. 2 (c. 1825), Wellcome MS 1590. For his experiments on soap, see Wellcome MS 3802.

74. J. Percy, *The metallurgy of lead, including desilverization and cupellation* (London, John Murray, 1870), p. 122.

75. H.L. Pattinson to J.B. Anderson, 11 December 1829.

76. H.L. Pattinson, 'A Copy of Two Discourses on Chemistry Composed in January 1819', First Discourse, Wellcome MS 3801.

77. H.L. Pattinson to Ralph [ ], 1 June 1828, Pattinson Letter Book, vol. 1.

78. H.L. Pattinson to J.M. Greenhow, 4 April 1838, Pattinson Letter Book vol. 3, Wellcome MS 3807.

79. J.R.R. Christie, 'Scottish scientific community'.

80. R.G.W. Anderson, *The Playfair Collection and the teaching of chemistry at Edinburgh, 1713-1858* (Edinburgh, The Royal Scottish Museum, 1978).

81. Joseph Black, *Lectures on the elements of chemistry*, ed. John Robison (2 vols. Edinburgh, W. Creech, 1803), vol.1, pp. 4-5.

82. J.B. Morrell, 'Thomas Thomson: Professor of Chemistry and University reformer', *Brit. J. Hist. Sci.*, **4** (1968-9), 245-65; and *idem*, 'The chemist breeders: The research schools of Liebig and Thomas Thomson', *Ambix*, **19** (1972), 1-46.

83. A. Kent, 'The Royal Philosophical Society of Glasgow, a

sesquicentennial address', *Phil. J.*, **4** (1967), p. 45. For examples of its activities in the 1830s see also Jack Morrell and Arnold Thackray, *Gentlemen of science: Early years of the British Association for the Advancement of Science* (Oxford, Clarendon Press, 1981), pp. 202-4.

84. W.V. Farrar, 'Andrew Ure, FRS and the philosophy of manufactures', *Notes Recs. Roy. Soc.*, **27** (1972-3), 299-324.

85. See A. and N.L. Clow, *The Chemical Revolution*. See also J.B. Morrell, 'Reflections on the history of Scottish science', *Hist. Sci.*, **12** (1974), 81-94.

86. E. M. Melhado, *Jacob Berzelius: The emergence of his chemical system* (Madison, Wis., University of Wisconsin Press, 1982). See also C.A. Russell, 'Introduction, commentary and notes' to J.J. Berzelius, *Essai sur Theorie des Proportions Chimiques et sur l'influence chimique de l'Electricite* (New York, Johnson Reprint Corporation, 1972), first ed. 1819; *idem*, 'The electrochemical theory of Berzelius' I – 'Origins of the theory', *Ann. Sci.*, **19** (1963), 117-26; *idem*, II -'An electrochemical view of matter', *ibid.*, 127-45; *idem*, 'Berzelius and the atomic theory' in D.S.L. Cardwell, ed., *John Dalton and the progress of science* (Manchester, Manchester University Press, 1968).

87. Maurice P. Crosland, *Historical studies in the language of chemistry* (London, Heinemann, 1962), pp. 265-81.

88. On Thomson's chemistry see J. R. Partington, 'Thomas Thomson, 1773-1852', *Ann. Sci.*, **6** (1948-50), 115-26.

89. Thomas Thomson, *An attempt to establish the first principles of chemistry by experiment* (2 vols., London, 1825).

90. *Idem, An outline of the sciences of heat and electricity* (Edinburgh, William Blackwood, 1830), p. ix.

91. T. Thomson to R. Jameson, 9 September 1817, quoted in J.B. Morrell, 'Thomas Thomson', p. 249.

92. Royal Society of London, *Catalogue of scientific papers, 1800-1863* (London, HMSO, 1867).

93. *Professors' Fees Books*, Record Office, University College London. On the purchasing power of Victorian incomes, £300 would sustain a servant, see Patricia Branca, *Silent sisterhood: Middle-class women in the Victorian home* (London, Croom Helm, 1975), pp. 54-5.

94. Graham's intellectual debt to Thomson is made clear in Michael Dennis Swords, 'The chemical philosophy of Thomas Graham', Ph.D Dissertation, Case Western Reserve University, 1973.

95. R.L. Ziemacki, 'Humphry Davy and the conflict of traditions in early nineteenth-century British chemistry', Ph.D Dissertation, Cambridge University, 1977, emphasises Davy's debt to eighteenth- century English chemical philosphy.

96. 'Chemistry', *Penny Cyclopedia* (London, Society for the Diffusion of Useful Knowledge, 1837), vol. 7, pp. 31-8. This anonymously authored article is attributed to Phillips in his obituary in *Gent. Mag.*, n.s. **36** (1851), 208.

97. William Whewell portrayed Faraday as the scientist who had laid the true groundwork for what had merely been speculation by Berzelius. W. Whewell, *History of the inductive sciences from the earliest to the present time* (3 vols. London, Parker, 1837), vol. 3, pp. 154-76.

98. For a detailed study of the fundamental differences between the approaches of metropolitan chemists and Berzelius, see Trevor H. Levere, *Affinity and matter: Elements of chemical philosophy, 1800-65* (Oxford, Clarendon Press, 1971). See also L. Pearce Williams, *Michael Faraday: A biography* (New York, Simon and Schuster, 1971) and David M. Knight, *The transcendental part of chemistry* (Folkstone, Dawson, 1978).

99. It is unclear just how important Boscovitch's model was to Davy and Faraday. L. Pearce Williams has emphasised its importance, 'Boscovitch and the British chemists', in *Roger Joseph Boscovitch: Studies of his life and work on the 250th anniversary of his birth*, ed. Lancelot Law Whyte (London, George Allen and Unwin, 1961), pp. 153-67. P.M. Heimann has challenged this interpretation in 'Faraday's theories of matter and electricity', *Brit. J. Hist. Sci.*, **5** (1970-1), 235-57.

100. William Whewell, 'On the employment of notation in chemistry', *J. Roy. Inst.*, **1** (1831), p. 453.

101. M. Faraday to W. Whewell, 21 February 1831, in L. Pearce Williams, *The selected correspondence of Michael Faraday* (2 vols., Cambridge, Cambridge University Press, 1971), vol. 1, p. 190.

102. J.F. Daniell, *An introduction to the study of chemical philosophy, being a preparatory view of the forces which concur to the production of chemical phenomena* (London, B. Fellowes, 1839), p. 504.

103. Graham's textbook can be dated from a review in *The Lancet*, **2** (1838-39), [23 March 1839], 20-4. Part 1 appeared in 1837 and Part 2 in 1838. The book is usually given the date 1842.

104. [Thomas Thomson], 'History and present state of chemical science', *Edin. Rev.*, **50** (1829), 275-6; attributed to Thomson, *Wellesley Index*, vol. 1, p. 471.

105. Both quotations from Francis Lunn, 'Chemistry', *Encyclopedia Metropolitana* (London, B. Fellowes, 1830), vol. 4, p. 596.

106. J.F.W. Herschel, 'Sound', *Encyclopedia Metropolitana*, vol. 4, p. 810. For Herschel's role in the introduction of Leibnizian notation see J.M. Dubbey, 'The introduction of the differential notation to Great Britain', *Ann. Sci*, **19** (1963), 37-48. See also J.F.W. Herschel, *A preliminary discourse on the study of natural philosophy* (London, Longman, Rees, Orm, Brown, and Green, 1830); David B. Wilson, 'Herschel and Whewell's version of Newtonianism', *J. Hist. Ideas*, **35** (1974), 79-97.

107. Charles Babbage, *Reflections on the decline of science in England and on some of its causes* (London, B. Fellowes, 1830). For the famous controversies over this book see Nathan Reingold, 'Babbage and Moll on the state of science in Great Britain', *Brit. J. Hist. Sci.*, **4** (1968), 58-64. See also Susan Faye Cannon, *Science in culture* (Folkstone, Dawson, 1978), pp. 167-200. For a more recent treatment see Jack Morrell and Arnold Thackray, *Gentlemen of science*.

108 'Decline of science in England', *Quart. Rev.*, **43** (1830), 305-42 [attributed to Brewster by the *Wellesley Index*, vol. 1, p. 710]. Brewster's auto-review is 'Observations on the decline of science in England', *Edin. J. Sci.*, n.s. **5** (1831), 1-16. He stressed chemistry particularly in 'Chemical science in England', *ibid.*, **6** (1832), 100-7.

109. [G. Moll], *On the alleged decline of science in England* (London, 1831).

110. J.F. Daniell, *An introductory lecture delivered in King's College London, 11 October 1831* (London, B. Fellowes, 1831).

111. On the BAAS see A.D. Orange, 'The origin of the British Association for the Advancement of Science', *Brit. J. Hist. Sci.*, **6** (1972), 152-76; O.J.R. Howarth, *The British Association for the Advancement of Science: A retrospect, 1831-1921* (London, BAAS, 1921); and Jack Morrell and Arnold Thackray, *Gentlemen of Science*. See also S.F. Cannon, *Science in culture*, pp. 201-24.

112. For the notion of 'gentlemen of science' see Jack Morrell and Arnold Thackray, *Gentlemen of Science*, chapter 1.

113. 'Chemical Committee', 'Recommendations of the Subcommittees', in BAAS, *Reports* (1831), p. 53.

114. 'Committee for Chemistry, &c.', 'Recommendations of the Committees', in BAAS, *Reports* (1832), p. 116.

115. 'Fourth Meeting of the British Association for the Advancement of Science', *Athenaeum*, 20 September 1834, p. 698.

116. 'Report of the Committee on Chemical Notation', in 'Reports on the State of Science', in BAAS, *Reports* (1835), p. 207.

117. T. Graham, 'Address to Section B', *Athenaeum*, 31 August 1839, p. 644.

Chapter two

1. Karl Hufbauer, *The formation of the German chemical community* (Berkeley, Calif., University of California Press, 1982).

2. B.H. Gustin, 'The emergence of the German chemical profession, 1790-1867', PhD Dissertation, University of Chicago, 1975. See also W.V. Farrar, 'Science and the German university system', in M. Crosland, ed., *The development of science in Western Europe* (London, Macmillan, 1975), pp. 193-8. See also J.B. Morrell, 'The chemist breeders: The research schools of Liebig and Thomas Thomson', *Ambix*, **19** (1972), 1-42; W.H. Brock, 'Liebigiana: Old and new perspectives', *Hist. Sci.*, **19** (1981), 201-18. For a transatlantic perspective see Owen Hannaway, 'The German model of chemical education in America: Ira Remsen at Johns Hopkins', *Ambix*, **23** (1976), 145-64.

3. Arnim Wankmüller, 'Ausländische Studierende der Pharmazie und Chemie bei Liebig in Giessen', *Tübinger Apothekensgeschichtliche Abhandlungen* (Stuttgart, Deutscher Apotheker Verlag, 1967), **15**.

4. [Chemical Society of London], *Jubilee of the Chemical Society of London: Record of the proceedings together with an account of the history and development of the Society, 1841-1891* (London, Chemical Society, 1896), p. 117.

5. 'Chemical Society of London', *Inventors' Advocate and Journal of Industry*, 27 February 1841, 140-1; and 'Chemical Society of London', *Polytechnic J.*, **4** (1841), pp. 253-4. These two journals reported the first meeting in almost identical terms.

6. Society of Apothecaries, 'Minutes of the Court of Assistants', **12** (1834-45), pp. 60-1, MS 8200, Guildhall, London. By 1839 the Museum had not yet been completed, *ibid.*, p. 294.

7. See *Jubilee of the Chemical Society*, pp. 119-21; and [Chemical Society of London], *History, constitution and laws of the Chemical Society of London* (London, Richard and John E. Taylor, 1842), p. 4.

8. 'Chemical Society of London', *Inventors' Advocate and Journal of Industry*, 3 April 1841, p. 220.

9. *Jubilee of the Chemical Society*, pp. 118-20, lists the founders.

10. Gardner is in the *DNB*, which lists him as a Licenciate of the Society of Apothecaries, but there is no record of his application in the archives of the Society in the Guildhall Library, London. He did eventually obtain a Giessen M.D., *The London and provincial medical directory* (London, John Churchill, 1850). However, the Giessen M.D. was notoriously 'purchasable' for £22 at the time, *The medical directory for Great Britain and Ireland for 1845*, comp. and ed. a Country Surgeon and General Practitioner (London, Medical Directory Office, 1845), pp. 665- 6. See also [Jacob Bell], 'The Rise and Progress of a Philosopher', *Pharm. J.*, **6** (1846-7), 141. [G.L.M. Strauss], *Reminiscences of an old Bohemian* (new ed.; London, Tinsley Brothers, 1883), p. 268, mentions that Gardner worked as German-English translator for *The Lancet* during the early 1840s. He also did translations for Liebig, whom he and Bullock both met, B. Lespius, ed., 'Festschrift zur Feier des 50. jährigen Bestehen der Deutschen Chemischen Gesellschaft und des 100. Geburtstages ihres Begründers August Wilhelm von Hofmann', *Ber.*, **51** (1918), pt. 2, Sonderheft, photograph opposite p. 8. Indeed, Bullock studied in Liebig's laboratory during the winter semester of 1839, Wankmüller, 'Ausländische Studirende', p. 12. Bullock also spent some time studying in Paris and was listed in contemporary postal directories as a chemist and druggist and operative chemist. Little more is known about Bullock except that he was a member of the Chemical Society and that he must have been fairly prosperous with property and a practice in Mayfair during the early 1840s, 'The late Mr. Bullock', *Chem. & Drug.*, **66** (1905), 882-3, 886. For Jacob Bell's view of Bullock, see Galen, 'Passing events', *Pharm. Times*, 5 September 1846, pp. 17-18.

11. J. Lloyd Bullock, 'A lecture on the state of pharmacy in England and its importance to the public; with remarks on the Pharmaceutical Society', *Chemist*, **5** (1844), p. 282.

12. For Gardner and Bullock's attempt to link science directly to personal gain at the Royal College of Chemistry see 'Amorphous Quinine', *Pharm. J.*, **6** (1846-7), 160-72 and G.K. Roberts, 'The establishment of the Royal College of Chemistry: An investigation of the social context of early-Victorian chemistry', *Hist. Stud. Phys. Sci*, **7** (1976), p. 464.

13. William Gregory, *Letter to the Rt. Hon. George, Earl of Aberdeen on the state of the schools of chemistry in the United Kingdom* (London, Taylor and Walton, 1842).

14. *Ibid.*, pp. 24-6. Liebig's views were subsequently published in English; Justus Liebig, *Familiar letters on chemistry and its relation to commerce, physiology, and agriculture* (London, Taylor and Walton, 1843).

15. *Ibid.*, p. 3.

16. Gardner and Bullock's initial plan is detailed in [John Gardner and John Lloyd Bullock], 'For a Practical Chemical School', 1843, Royal

Institution Archives, London. See the discussion in G.K. Roberts, 'Royal College of Chemistry', pp. 460-4, and A.S. Forgan, 'The Royal Institution of Great Britain, 1840-1873', PhD Dissertation, University of London, 1977.

17. *Proposal for establishing a College of Chemistry for promoting the science and its application to agriculture, arts and medicine* (London, G.J. Palmer, [1844]), Add. MSS No. 40553, ff. 21-9, British Library, London. *To agriculturists: Supplement to the proposal for establishing a College of Chemistry* (London, n.d), and *Supplement to the proposal for establishing a College of Chemistry: To the proprietors of mines and metallurgists* (London, n.d). Both supplements can be dated, see Royal College of Chemistry, 'Minutes of the Council of the College, 1845-51', 5 March 1845 and 25 January 1845, Imperial College London, College Archives, C/3/3.

18. The negotiations over Hofmann's appointment are discussed in G.K. Roberts, 'The Royal College of Chemistry (1845-1853): A social history of chemistry in early-Victorian England', PhD Dissertation, The Johns Hopkins University, 1973, pp. 265-73. Financial considerations and career prospects featured in Hofmann's decision to take the post. He calculated that he could save enough from his projected English earnings to marry Liebig's niece within two years, and so was willing to risk a contract of that length at the untried institution provided that his existing post at Bonn would be held open for him during that time as insurance. A. Hofmann to J. Liebig, 26 June 1845, Liebig-Hofmann Correspondence, Bayerische Staatsbibliothek, Munich. See W.H. Brock, ed., *Justus von Liebig und August Wilhelm Hofmann in ihren Briefen 1841-1873* (Weinsheim, Verlag Chemie, itp). For details of Hofmann's contract, see RCC, 'Council Minutes', 24 and 25 September 1845. The Prince Consort's assistance in obtaining Hofmann's leave of absence from Bonn has often been documented, see, for example, Sir Patrick Linstead, *The Prince Consort and the founding of Imperial College* (London, Imperial College of Science and Technology, ˙1961). On Hofmann's appointment see also J. Bentley, 'The Chemical Department of the Royal School of Mines: Its origins and development under A.W. Hofmann', *Ambix*, **17** (1970), 153-81.

19. Royal College of Chemistry, 'Alphabetical List of Members, 1844-46', Imperial College London, College Archives, C/2; and Royal College of Chemistry, 'Subscribers and Donors, 1846-52', Imperial College London, College Archives, C/2/2. Support is analysed in Roberts, Dissertation, p. 194 and Appendix I.

20. Hugh Hale Bellot, *University College London, 1826-1926* (London, University of London Press, 1929), p. 125. See also J.B. Morrell, 'Practical chemistry in the University of Edinburgh', *Ambix*, **16** (1969), 69-73; and 'Evidence, oral and documentary taken before the Commissioners for visiting the Universities of Scotland with Appendix and Index – Edinburgh', *P.P.* 1837 [92] XXXV. 537.

21. H.H. Bellot, *University College*, chapter 7. The curriculum is described in University of London, 'Minutes of the first subcommittee of the Faculty of Medicine: Subcommittee to consider the course of study required of candidates for degrees in medicine who shall hereafter commence their

medical studies, 8 August 1837 to 12 October 1837', 22 August 1837, University of London, Senate House Library.

22. University College London, *Annual reports* (London, Richard Taylor, 1840), p. 19.

23. Thomas Wakley (1795-1862), radical politician, medical reformer, and editor of the polemical medical journal *The Lancet* took up this theme. Wakley was one of Liebig's earliest English promoters, publishing his university lectures on organic chemistry and supplementary material on its applications quite early in the 1840s. See, for example, 'Review of *Elementary instruction in chemical analysis...*', *The Lancet*, 21 October 1843, pp. 101-2; 'Attendance on Liebig's class', *ibid.*, 27 January 1844, p. 591; 'The laboratory at Giessen', *ibid.*, 18 May 1844, p. 261; 'Remarks on the new College of Chemistry', *ibid.*, 7 September 1844, p. 736. 'The science of chemistry', *ibid.*, 16 November 1844, pp. 231-2, and 'The College of Chemistry', *ibid.*, 11 October 1845, pp. 403-4.

24. Jacob Bell and Theophilus Redwood, *Historical sketch of the progress of pharmacy in Great Britain* (London, Pharmaceutical Society, 1880), pp. 80, 91. S.W.F. Holloway, 'The Apothecaries' Act of 1815: A reinterpretation', *Med. Hist.*, **10** (1966), p. 125, and *idem.*, 'Medical education in England, 1830-1858: A sociological analysis', *History*, **49** (1964), pp. 311-12. See also 'Pharmaceutical Meetings', *Pharm. J.*, **1** (1841), 6-7, and Jacob Bell, 'General observations by the editor', *ibid.*, 38-9.

25. See Chapter 1 above and S.F. Gray, *The operative chemist: being a practical display of the arts and manufactures which depend upon chemical principles* (London, Hurst, Chance & Co., 1828). See also Jacob Bell, *Observations addressed to the chemists and druggists of Great Britain on the Pharmaceutical Society* (London, C. Whiting, 1841), p. 6. Competition was a very real issue amongst chemists and druggists at the time. In Sheffield, during the period 1801-1841, the number of chemists and druggists quintupled while the population only doubled. Chemists and druggists tended to diversify in order to cope with the competition, R.F. Bud, 'The discipline of chemistry: The origins and early years of the Chemical Society of London', PhD Dissertation, University of Pennsylvania, 1980.

26. Jacob Bell and Theophilus Redwood, *Historical sketch*, pp. 87-8.

27. Jacob Bell, 'Pharmaceutical meetings', pp. 4-5; *idem*, 'On the constitution of the "Pharmaceutical Society of Great Britain"', *Pharm. J.*, **1** (1841), 4-13.

28. Jacob Bell, *Observations*, p. 10; T.E. Wallis, *History of the School of Pharmacy, University of London* (London, The Pharmaceutical Press, 1964), p. 3; Joseph Ince, 'The history of the School of Pharmacy', *Pharm. J.* (1903), pt.1, p. 282.

29. J. Lloyd Bullock, 'A lecture on the state of pharmacy in England and its importance to the public with remarks on the Pharmaceutical Society', *The Chemist*, **5** (1844), 277-82. See also editorial comments under various titles, *ibid.*, pp. 126-7, 144, 178, 226-7, 323-4, 555.

30. J.A. Scott-Watson, *The history of the Royal Agricultural Society of England, 1839-1939* (London, The Royal Agricultural Society, 1939), pp. 18-19.

31. F.M.L. Thompson, 'The Second Agricultural Revolution, 1815-1880', *Econ. Hist. Rev.*, 2nd s, **21** (1968), 62-77. On the complex political and social context of high farming see D.C. Moore, 'The Corn Laws and High Farming', *ibid.*, 2nd s, **18** (1965), 544-54.

32. F.M.L. Thompson, 'Second Agricultural Revolution', p. 71.

33. 'Tables of the revenue, population and commerce of the United Kingdom and its Dependencies Part XX. Sect. A, 1850', *P.P.* 1852 [1466 and 1466-1] LII. 1, p. 215. On guano see W.M. Mathew, *The House of Gibbs and the Peruvian guano monopoly*, Royal Historical Society Studies in History no. 25 (London, 1981).

34. David Spring, 'The English landed estate in the age of coal and iron, 1830-1880', *J. Econ. Hist.*, **11** (1951), 3-24. *The mining guide, containing particulars of each mine, British and foreign, its situation and produce* (London, Mining Journal, 1853), pp. iii-iv, notes that the number of English mining companies known in the market grew from under fifty to about 520 during the decade after 1843.

35. 'Prospectus for the Museum of Economic Geology', Commissioners of Woods and Forests to H.M. Treasury, Great Britain, Public Record Office: Treasury Papers, T. 1-3776. See also Margaret Reeks, *Register of the Associates and Old Students of the Royal School of Mines and the history of the Royal School of Mines* (London, Royal School of Mines Old Students Association, 1920), pp. 15-16.

36. Accounts can be found in Mr. Chawner to Commissioners of Woods and Forests, 20 September 1839, GB, PRO, T. 1-3776. For details of Phillips' earnings R. Phillips to H. De la Beche, 9 January 1841, Entry Book of In and Out Letters, 1835-42, pp. 180-82, MS GSM1/1, Institute of Geological Sciences, London; and 'Revised regulations respecting the office of curator of the Museum of Economic Geology', H. De la Beche to R. Phillips, 3 February 1841, *ibid.*, pp. 191-4.

37. See 'Prospectus for the Museum of Economic Geology' and Reeks, *Register*, p. 15.

38. Justus Liebig, *Organic chemistry in its applications to agriculture and physiology*, ed. Lyon Playfair (London, Taylor and Walton, 1840).

39. *Ibid.*, pp. 138, 140.

40. This was noted ironically in Thomas Graham to Mrs. J. Reid, 29 September 1844, in Robert Angus Smith, *The life and works of Thomas Graham, D.C.L., F.R.S.* (Glasgow, John Smith and Sons, 1884), p. 44. The BAAS decision is recorded in British Association, *Reports* (1844), p. xxiii. In the event, the Royal Agricultural Society agreed and voted £350 for the project; see Royal Agricultural Society, 'Council Minutes', 5 February 1845, Royal Agricultural Society Archives.

41. Alexander Ramsay, *History of the Highland and Agricultural Society of Scotland* (Edinburgh, William Blackwood and Sons, 1879). The Society's views on analysis can be traced in Highland and Agricultural Society, 'Sederunt Books', 10 February 1842, pp. 128-9; 8 March 1842, p. 152; 22 March 1842, pp. 163-4 and 5 July 1842, pp. 268-9, Royal Highland and Agricultural Society, Edinburgh.

42. 'Proceedings of the Agricultural Chemistry Association of Scotland, 1845', *Trans. Highland Ag. Soc.*, 3rd, **1** (1843-5), 459-69.

43. For the appointment of Johnston see 'Meeting of agriculturalists', *Scotsman*, 5 July 1843. On the Association's activities see 'Report of the Committee of Management of the Agricultural Chemistry Association of Scotland', *J. Ag.*, **1** (1845-7), 230-62; and David Milne, *The Highland and Agricultural Society and the Agricultural Chemistry Association; respective plans for promoting scientific agriculture* (Edinburgh, n. pub., 1848).

44. 'Carrickfergus Agricultural Society', *Pharm. Times*, **2** (1847), 52-3. See also 'Report of the Committee of Management of the Agricultural Chemistry Association of Scotland', 260-1.

45. 'Royal Agricultural and Commercial Society of British Guiana', in 'Fourth report of the Select Committee on sugar and coffee planting: Together with the minutes of evidence and appendix', *P.P.* 1847-48 (184) XXIII.II. 1, pp. 62-7. See also 'On the duties of an agricultural chemist located at one of our sugar producing Colonies (Abstract from a report ... presented to the Governor of British Guiana by John Shier, LL.D., Agricultural Chemist to the Colony)', *Pharm. J.*, **7** (1847-8), 134-7. On Shier (1807-1854), who was educated at Marischal College, Aberdeen, and under D.B. Reid in Edinburgh, see Alexander Findlay, *The teaching of chemistry in the Universities of Aberdeen*, Aberdeen University Studies 12 (Aberdeen, The University Press, 1935), pp. 36-9. On conditions in British Guiana at mid-century see Barton Premium, *Eight years in British Guiana: Being the journal of a resident in the Province from 1840 to 1848 ...* (London, Longman, Brown, Green, and Longman, 1850); J.R. Mandle, *The plantation economy, population and economic change in Guyana 1835-1960* (Philadelphia, Temple University Press, 1973), pp. 17-25; and R.E.G. Farley, 'Aspects of the economic history of British Guiana, 1781-1852: A study of economic and social change on the Southern Carribean Frontier', PhD Dissertation, University of London, 1956.

46. 'Report of the Committee of Management of the Agricultural Chemistry Association of Scotland', p. 261.

47. 'As soon as the Corn-law question comes on for the discussion of parliament, the blow will be struck for the establishment of Agricultural Colleges, and your capital letter or rather the extract of it referring to the necessity of regenerating chemistry in England will privately be brought under the notice of our premier Sir Robert Peel by some of my friends at Court'. (L. Playfair to J. Liebig, 27 December 1841, Liebigiana 58, Bayerische Staatsbibliothek, Munich). Earl of Clarendon, 'The Royal College of Chemistry', *London Medical Gazette*, **2** (1846), p. 1011.

48. Charles G.B. Daubeny, 'Lecture on the application of science to agriculture, 9 December 1841', *J. Roy. Ag. Soc.*, **3** (1842), 136-57; *idem*, 'On public institutions for the advancement of agricultural science which exist in other countries and on plans which have been set on foot by individuals with a similar intent in our own', *ibid.*, 364-86. See also 'Lecture on institutions for the improvement of agriculture', 27 April 1842, Sherard MSS, No. 278, Oxford University, Bodleian Library.

206   *Notes to chapter two*

49. 'Agricultural College', *Eng. J. Ed.*, **1** (1843), 383-4. See also, Royal Agricultural College, *Prospectus* (Cirencester, n. pub., 1846).

50. Charles G.B. Daubeny, *A Lecture on the institutions for the better education of the farming classes, especially with reference to the proposed Agricultural College near Cirencester, 14 May 1844* (Oxford, T. Combe, 1844), pp. 11, 27.

51. [Royal College of Chemistry], *Supplement to agriculturists*, p. 8.

52. On agriculturists, see Daubeny, *Lecture on institutions, 1844*, p. 9; and William Stark, *A letter to the Rt. Hon. Lord Wodehouse, President of the Norfolk Agricultural Association on the use of chemical manures* (Norwich, Bacon & Co., 1844), p. 7. On chemists and druggists, see *The Chemist*, **2** (1844), p. 2.

53. James Young to his mother, 19 October 1838, Strathclyde University Archives, James Young Papers (T-YOU).

54. Edmund Potter, *'Calico printing as an art manufacture': A lecture read before the Society of Arts, 22 April 1852* (London, John Chapman, 1852); 'Report from the Select Committee on the copyright of designs together with the minutes of evidence taken before them and an appendix and index', *P.P.* 1840 (442) VI. 1, p. 494. On chemistry and calico printing before synthetic dyestuffs see Edward Baines, *History of the cotton manufacture in Great Britain* (London, Fisher, Fisher and Jackson, 1835), pp. 245-85; Geoffrey Turnbull, *A history of the calico printing industry of Great Britain* (Altrincham, John Sherratt and Son, 1951); P. Floud, 'The English contribution to the chemsitry of calico printing before Perkin', *CIBA Review*, 1961, pt. 1, pp. 8-14; and C.M. Mellor and D.S.L. Cardwell, 'Dyes and dyeing, 1775-1860', *Brit. J. Hist. Sci.*, **1** (1963), 265-79. Some firms had quite elaborate laboratories, 'A day in a Lancashire print works', *Penny Magazine*, **12** (1843), 265-79.

55. 'Advertisement', *The Chemist*, **1** (1840), 1-4. The full title of the journal was *The Chemist, or Reporter of Chemical Discoveries and Improvements, and Protector of the Rights of the Chemist and Chemical Manufacturer*.

56. 'Introduction', *The Chemist*, **1** (1840), p. iii.

57. 'Advertisement', *The Chemist*, **5** (1844), p. 2.

58. [Royal College of Chemistry], *Supplement to the proprietors of mines*, pp. 4, 7; and J. H. Morris and L.J. Williams, *The South Wales coal industry, 1841-1875* (Cardiff, University of Wales Press, 1958). For an assessment of the contemporary state of mining science, see Roy Porter, 'The Industrial Revolution and the rise of geology', in Mikulás Teich and Robert Young, eds., *Changing perspectives in the history of science: Essays in honour of Joseph Needham* (London, Heinemann, 1973), pp. 320-43.

59. O.O.G.M. MacDonagh, 'Coal mines regulation: The first decade, 1842-1852', in R. Robson ed., *Ideas and institutions of Victorian Britain* (London, G. Bell & Co, 1967); Neil K. Buxton, *The economic development of the coal industry, from Industrial Revolution to the present day* (London, Batsford Academic, 1978), chapter 6.2.

60. Dean Chandler and A. Douglas Lacey, *The rise of the gas industry in Britain* (London, The British Gas Council, 1949), pp. 73-6.

61. 'City Commissioners of Sewers', *Patent J.*, **4** (1847-8), 259-61.

62. 'Annual Report', *Min. Proc. Inst. Civil Eng.*, **4** (1845), p. 6. This was

of course analogous to arguments about medical professionalization. See Philip Elliot, *The sociology of the professions* (London, Macmillan, 1972). The 1841 Census counted 959 civil engineers; the 1851 Census counted 3,009. 'Census of Great Britain, 1841. Abstract of the Answers and Returns: Occupation Abstract, pt. 1 – England, Wales and Islands in the British Seas', *P.P.* 1844 (587) XXVII. 1; 'pt. 2 – Scotland', *P.P.* 1844 (588) XXVII. 385. 'Census of Great Britain, 1851. Population Tables II. Ages, Civil Condition, Occupations and Birthplace of the People', *P.P.* 1852-53 [1691-I] LXXXVIII (I). 1, Table 54, 'Classified Arrangement of the Occupations of the People in 1851', pp. cxxviii-cxlix.

63. On these courses, see respectively University of Durham, *Calendar* (Durham, 1839 onwards). C.E. Whiting, *The University of Durham, 1832-1932* (London, Sheldon Press, 1932). King's College London, *Calendar* (London, B. Fellowes, 1840 onwards); King's College London, *Report from the Council to the Annual General Court of Governors and Proprietors, Monday 28 April 1845* (London, Rich. Clay, 1845), p. 25; F.J.C. Hearnshaw, *The centenary history of King's College London* (London, G. Harrap, 1929); and D.H. Hey, 'Schools of chemistry in Great Britain and Ireland: XVIII – King's College London', *J. Roy. Inst. Chem.*, **79** (1955), 305-15. Putney College for Civil Engineers, *College for Civil Engineers and of general practical and scientific training* (Putney, n. pub., 1845); W.R. Howard, 'Presidential address', *Trans. Soc. Eng.*, **45** (1954), p. 9.

64. [Royal College of Chemistry], *Proposal*, p. 6.

65. *Idem, Supplement to the proprietors of mines*, p. 1.

66. G.K. Roberts, 'Royal College of Chemistry', pp. 473-5, summarises support for the College. For more detail, see *idem*, Dissertation, pp. 194-264 and Appendix 1.

67. C.A. Russell with N.G. Coley and G.K. Roberts, *Chemists by profession: The origins and rise of the Royal Institute of Chemistry* (Milton Keynes, Open University Press and Royal Institute of Chemistry, 1977), pp. 78-81; and G.K. Roberts, Dissertation, pp. 314-47 and Appendix 2.

68. G.K. Roberts, Dissertation, pp. 360-9. In fact, most of the 'problems' submitted were analytical in nature; Hofmann's report to Council in 1846 indicated the range: quinine, dyestuffs, coal tar products, foodstuffs, soils, plant ashes, gun metals, alloys, minerals, glass and spa waters, Royal College of Chemistry, 'Council Minutes', 12 August 1846.

69. University College London, 'Council Minutes', 8 February 1845; and University College London, 'Committee of Management Minutes', 24 September 1845, University College London, Record Office. King's College London, 'Minutes of Council', 1 August 1845, King's College London, Library.

70. On W.A. Miller's appointment at King's, see J.F.C. Hearnshaw, *Centenary history of King's College London*, p. 174. On the appointment of George Fownes at University College, see University College London, 'Council Minutes', 5 February 1845.

71. 'Birkbeck Testimonial', ff. A46 B1R, College Archives, University College London.

72. Thomas George Tilley, *An inaugural lecture on chemistry … read in the*

Queen's College Birmingham, ... *7 October 1844* (London, John Van Vossert, n.d), p. 20.

73. J. Sheridan Muspratt, 'Dr. Muspratt and Mr. Spencer', *Liverpool Mercury,* 19 April 1850; see also Gordon W. Roderick and Michael D. Stephens, *Scientific and technical education in nineteenth-century England: A symposium* (New York, Barnes and Noble, 1977), pp. 65-73.

74. Roughly 10% of the 356 students from 1845-1853 held an academic post at some stage in their careers at some thirty-eight institutions plus schools. G.K. Roberts, Dissertation, pp. 344-5; see also n. 68 above.

75. T.W. Moody and J.C. Beckett, *Queen's Belfast, 1845-1949: The history of the University* (London, Faber and Faber for the Queen's University Belfast, 1959). See also 'Queen's College Commission; Report of Her Majesty's Commissioners appointed to inquire into the progress and condition of the Queen's Colleges at Belfast, Cork and Galway; with minutes of evidence, documents and tables of returns', *P.P.* 1857-58 [2413] XXI. 53. The testimony of Thomas Andrews, Professor of Chemistry at Belfast, indicates that most chemistry students were from the medical, engineering, and agricultural faculties.

76. J.A. Richey, *Selections from educational records,* Bureau of Education, India, Part II, 1840-59 (Calcutta, Superintendent of Government Printing, India, 1922), p. 330. For scientific instruction in India see H.J.C. Larwood, 'Science in India before 1850', *Brit. J. Ed. Studs.,* 7 (1958-9), 36-49.

77. On the development of chemistry in Canada see W. Lash Miller, 'The beginnings of chemistry in Canada', in H.M. Tory, ed., *The history of science in Canada* (Toronto, Ryerson Press, 1939), pp. 21-34.

78. R. Steven Turner, 'The growth of professorial research in Prussia, 1818-1848', *Hist. Stud. Phys. Sci.,* 3 (1972), 137-82. See also B.H. Gustin, 'The emergence of the chemical profession in Germany'.

79. Thomas Graham to his mother, 12 November 1827 in R.A. Smith, *The life and works of Thomas Graham,* p. 283.

80. 'Report of the Committee Appointed by the Senate to Examine the Testimonials for the Chair of Chemistry, 3 June 1837', College Archives, University College London.

81. 'Report of Senate on the Applications and Testimonials of Candidates for the Prof^r of Practical Chem^y, 1849', College Archives, University College London.

82. L. Playfair to J. Liebig, 4 June 1842, Liebigiana 58, no. 10, Bayerische Staatsbibliothek, Munich.

83. J. Mercer to L. Playfair, 18 January 1843, Mercer-Playfair Correspondence, Manchester Central Library.

84. [Edward Frankland], *Sketches from the life of Edward Frankland* (London, privately pub., 1901), pp. 271-2.

85. [The Chemical Society], *Jubilee of the Chemical Society: Record of the proceedings together with an account of the history and development of the Society, 1841-1891* (London, The Chemical Society, 1896), p. 197.

86. 'Report', *Mem. Proc. Chem. Soc.,* 3 (1845-7), p. 347. See also The Chemical Society, 'Minutes of Council', 19 April 1847, p. 29, Royal Society of Chemistry.

87. A.W. Hofmann to J. Liebig, 15 January 1846 and 7 February 1847,

Hofmann-Liebig Correspondence. G.K. Roberts, Dissertation, p. 335, calculates that Hofmann and his pre-1853 pupils contributed approximately one-fifth of all *Quart. J. Chem. Soc.* papers up to the time Hofmann left England in 1865. Adding in contributions from post-1853 students makes the college's influence on the journal in this period even more pronounced.

88. The Society's publishing performance is analysed in detail in R.F. Bud, Dissertation, pp. 317- 47.

89. See for example, 'On the composition of the fire-damp of the Newcastle coal mines', *Mem. Proc. Chem. Soc.*, **3** (1845-7), 7-10; 'On the supply of iodine from the kelp of Guernsey', *ibid.*, 252; and 'Note on the useful application of the refuse lime of gas-works', *ibid.*, **2** (1843-5), 358-60.

90. Royal College of Chemistry, *Reports of the Royal College of Chemistry and researches conducted in the laboratories in the years 1845-6-7* (London, 1849); 2nd vol. *For the years 1848-49-50-51* (London, 1853). In the main these papers were reprinted from the Chemical Society. On spa waters, see E.G. Schweitzer, 'Analysis of the Bonnington water near Leith, Scotland', *Mem. Proc. Chem. Soc.*, **2** (1843-5), 201-18; F.A. Abel and T.H. Rowney, 'On the mineral waters of Cheltenham', *Quart. J. Chem. Soc.*, **1** (1849), 193-212; G. Merck and R. Galloway, 'Analysis of a medicinal water from the neighbourhood of Bristol', *ibid.*, **2** (1850), 200-11; T.J. Herapath and W. Herapath, 'The waters of the Dead Sea', *ibid.*, 336-44.

On homologous series (and alkaloid investigations), the pattern was that Hofmann would do a 'leading' paper setting out a central problem and methods for dealing with it which students could then imitate and branch out from. For example, J.S. Muspratt and A.W. Hofmann, 'On toluidine, a new organic base', *Mem. Proc. Chem. Soc.*, **2** (1843-5), 367-83 was one of these leading papers. There followed from it a range of work by students: E.C. Nicholson, 'On cumidine, A new organic base', *Quart. J. Chem. Soc.*, **1** (1849), 2-11; G. Maule, 'On nitromesidine, a new organic base', *ibid.*, **2** (1850), 116-21; H.M. Noad, 'On the action of nitric acid on cymol', *Mem. Proc. Chem. Soc.*, **3** (1845-7), 421-39; F.A. Abel, 'On some products of the oxidation of cumol by nitric acid', *ibid.*, 441-7; F. Field, 'On the products of the decomposition of cuminate of ammonia by the action of heat', *ibid.*, 401-12; H. Medlock, 'Researches on the amyl series', *Quart. J. Chem. Soc.*, **1** (1849), 368-79, *ibid.*, **2** (1850), 212-15.

91. J.S. Muspratt and A.W. Hofmann, 'On toluidine', p. 368.

92. 'Liverpool College of Practical Chemistry under the Superintendence of Dr. Sheridan Muspratt: Prospectus with testimonials', Mus. 4.5, Record Office, Liverpool, Liverpool City Libraries.

93. 'Diary of Edward Frankland', 7 February 1848, p. 14, Archives, Royal Society, London. E. Frankland and H. Kolbe, 'Upon the chemical composition of metacetonic acid and some other bodies related to it', *Mem. Proc. Chem. Soc.*, **3** (1845-7), 386-91 (386-7). See also *idem*, 'On the product of the action of potassium on cyanide of ethyl', *Quart. J. Chem. Soc.*, **1** (1849), 60-74; H. Kolbe, 'Research on the electrolysis of organic compounds', *ibid.*, **2** (1850), 152-84; Edward Frankland, 'On the isolation of organic radicals', *ibid.*, 263-96; *idem*, 'Researches on the organic radicals: Part II, Amyl', *ibid.*, **3** (1851), 30-53.

94. R.F. Bud, Dissertation, p. 334, and Appendix 3.

95. J. Mercer, 'On the action of a mixture of red prussiate of potash and caustic alkali upon colouring matters', *Mem. Proc. Chem. Soc.*, **3** (1845-7), 320-1; Walter Crum, 'On the action of bleaching powder on the salts of copper and lead', *ibid.*, **2** (1843-5), 387-91.

96. C.O. Glassford and J. Napier, 'On the cyanides of the metals and their combinations with cyanide of potassium. Part 1, cyanide of gold', *Mem. Proc. Chem. Soc.*, **2** (1843-5), 82-92; 'Part 2, cyanide of silver', 92-7. J. Napier, 'Observations on the decomposition of metallic salts by an electric current', *ibid.*, 255-60; *idem*, 'Observations upon the decomposition of the double cyanides by an electric current', *ibid.*, 158-62; *idem*, 'On the unequal decomposition of electrolytes and the theory of electrolysis', *ibid.*, **3** (1845-7), 47-54; see also *idem*, 'On electrical endosmose', *ibid.*, 28-39.

97. The difference between specialty and technical criteria of significance is an enduring feature of modern science. For a description of the implications for industrial research, see David B. Herz, *The theory and practice of industrial research* (New York, McGraw-Hill, 1950), pp. 180-91.

98. H. Will, 'Observations on M. Reiset's remarks on the new method for the estimation of nitrogen in organic compounds, and also on the supposed part which the nitrogen of the atmosphere plays in the formation of ammonia', *Mem. Proc. Chem. Soc.*, **1** (1841-3), 197-208; *idem*, 'Extract from a letter from Dr. Will dated Giessen, November 10 1842', *ibid.*, p. 44. George Fownes, 'On the analysis of organic substances containing nitrogen', *ibid.*, 41-3. William Francis, 'Remarks on the determination of nitrogen in organic analysis', *ibid.*, 44-5.

99. For a detailed discussion, see R.F. Bud, Dissertation, pp. 339-43. Typical of this type of paper were various categories of water analyses; sixteen papers on this subject were published during the first nine years. In addition to analysing the active ingredients and quality of spa waters (n. 90 above), there were papers on the water supply of London written in the context of current debates about the quality of well water v. Thames water.

100. E.F. Teschemacher, 'On gun-cotton', *Mem. Proc. Chem. Soc.*, **3** (1845-7), 253-6; Robert Porrett and E.F. Teschemacher, 'On the chemical composition of gun-cotton', *ibid.*, 258-61; Robert Porrett, 'On the existence of a new vegeto-alkali in gun-cotton', *ibid.*, 287-90; 'Letter from J. Coathupe', *ibid.*, p. 329.

101. J.H. Gladstone, 'Contributions to the chemical history of gun-cotton and xyloidine', *Mem. Proc. Chem. Soc.*, **3** (1845-7), 412-21.

102. 'Charter of the Chemical Society', printed in *Jubilee of the Chemical Society*, pp. 135-6.

103. For a sociological interpretation of this type of discipline structure, see Richard Whitley, 'Umbrella and polytheistic scientific disciplines and their elites', *Soc. Stud. Sci.*, **6** (1976), 471-97.

Chapter three

1. John Gardner, 'An address delivered in the Royal College of Chemistry, 3 June 1846', *The Chemist* **7** (1846), 294-301.
2. Royal College of Chemistry, 'Minutes of the Council of the College,

1845-51', 12 August 1846, Imperial College London, College Archives, C/3/3.

3. For the contemporary definition of a liberal education see William Whewell, *On the principles of English university education* (London, John W. Parker, 1837). For a more recent treatment of the subject see Sheldon Rothblatt, *Tradition and change in English liberal education* (London, Faber & Faber, 1976). See also Robert G. McPherson, *Theories of higher education in nineteenth-century England* (Athens, Ga., U. of Georgia Press, 1959).

4. John Gardner, 'An address ... 3 June 1846', p. 296.

5. *Ibid.*

6. F. Wardle to C.C. Atkinson [1849], College Correspondence, College Archives, University College London.

7. Royal College of Chemistry, 'Council Minutes', 31 May 1848.

8. William Elliot to C.L. Bloxam, 3 October [1851], Bloxam Diaries, privately held by G.A. Bloxam. We are grateful to Mr. Bloxam for permission to make use of this material.

9. This view was promulgated by Hofmann himself; A.W. Hofmann, 'A page of scientific history: Reminiscences of the early days of the Royal College of Chemistry', *J. Sci.*, **8** (1871), 145-53. See also Sir F.A. Abel, 'The history of the Royal College of Chemistry and reminiscences of Hofmann's professorship' *J. Chem. Soc.*, **69** (1896), pt. 1, 580-96; Lyon Playfair, 'Personal reminiscences of Hofmann and of the conditions which led to the establishment of the Royal College of Chemistry and the appointment of its professor', *ibid.*, 575-9; and T.G. Chambers, *Register of the Associates and Old Students of the Royal College of Chemistry, the Royal School of Mines and the Royal College of Science with historical introduction and biographical notices and portraits of past and present professors* (London, Hazell, Watson & Viney, Ltd., 1896); and more recently John J. Beer, 'A.W. Hofmann and the founding of the Royal College of Chemistry', *J. Chem. Ed.*, **37** (1960), 248-51; Sir Patrick Linstead, *The Prince Consort and the founding of Imperial College* (London, Imperial College of Science and Technology, 1961); and J. Bentley, 'The Chemical Department of the Royal School of Mines. Its origins and development under A.W. Hofmann', *Ambix*, **17** (1970), p. 153.

10. A.W. Hofmann to J. Liebig, 5 April 1851. Liebig-Hofmann Correspondence, Bayerische Staatsbibliothek, Munich.

11. A.W. Hofmann, 20 March 1851, Liebig-Hofmann Correspondence.

12. R.A. Smith, 'The life and work of Thomas Graham, DCL, FRS, illustrated by sixty-four unpublished letters', ed. J.J. Coleman, *Proc. Phil. Soc. Glasgow*, **15** (1883-4), 319.

13. University College London, 'Professors' Fees Books', Record Office, University College London. The high proportion of chemistry students who subsequently became medical men is evident from comparing students in the class lists with *The London and provincial medical directory and general Medical Register* (London, John Churchill, five yearly intervals 1850-85).

14. T. Graham to C.C. Atkinson, 1 December 1838, College Correspondence, 4440, College Archives, University College London.

15. Edith Frame, 'Thomas Graham: A centenary account', *Phil. J.*, 7 (1970), 117-19.

16. T. Graham to C.C. Atkinson, 1 December 1838.

17. University of London, *Minutes of the Senate*, vol. 1: *March 4th 1837 to June 21st 1843* (London: Taylor and Francis, n.d.), 27 March 1839.

18. T. Graham to J. Graham, 29 October 1844, in R.A. Smith, 'Thomas Graham', p.450.

19. T. Graham to C.C. Atkinson, 1 December 1838.

20. University College London, *Calendar for the session 1853-54* (London, Richard Taylor, 1853).

21. J. Harris and W.H. Brock, 'From Giessen to Gower Street: Towards a biography of Alexander William Williamson (1824-1904)', *Ann. Sci.*, **31** (1974), 95-130.

22. G.K. Roberts, 'The Royal College of Chemistry (1845-53): A social history of chemistry in early-Victorian England', PhD Dissertation, The Johns Hopkins University, 1973. Table XIII lists the posts taken up by RCC students of the pre-1853 cohort.

23. We are grateful to Mr. Gordon Dyer, who is doing research on the British Army as a patron of science in the nineteenth century for bringing a statement of the criteria to our attention. F.A. Abel and C.L. Bloxam, *Hand-Book of Chemistry: Theoretical, practical and technical* (London, John Churchill, 1854).

24. See G.K. Roberts, Dissertation, Table XIII.

25. University of London, 'Committee appointed to consider the propriety of establishing a degree or degrees in science, and the conditions on which such degree or degrees should be conferred', in *Minutes of Committees, 1853-66*, University of London, Senate House Library, 18 May 1858, p. 72.

26. The memorialists were R. Murchison, C. Lyell, R. Owen, A. Ramsey, A. Farre, J. Hooker, G. Busk, P.M. Grey Egerton, W. Sharpey, W. Bowman, H.B. Jones, T.H. Huxley, Mr. Latham, J. Lubbock, J. Lindley, J. Beete Jukes, T. Bell, A. Henfrey, G.J. Allman. Bence Jones is a possible exception. He was a physician, but did have chemical interests and studied the subject at University College; *DNB*.

27. University of London, 'Committee to consider degrees in science', 7 July 1858, 19 January 1859, 13 April 1859, and 16 June 1859.

28. *Ibid.*, 8 July 1859, pp. 115-27.

29. H.B. Charlton, *Portrait of a University, 1851-1951: To commemorate the centenary of the University of Manchester*, 2nd ed. (Manchester, Manchester University Press, 1952), p. 28.

30. *Ibid.*, p. 30.

31. Joseph Thompson, *The Owens College: Its foundation and growth and its connection with the Victoria University, Manchester* (Manchester, Manchester University Press, 1886), p. 148.

32. Robert H. Kargon, *Science in Victorian Manchester: Enterprise and expertise* (Manchester, Manchester University Press, 1977), p. 164.

33. Owens College, 'Examination Papers', in *Calendar for the session 1851-52* (Manchester, T. Sowler & Sons, 1851), p. 26.

34. Owens College, 'Examination Papers' in *Calendar for the session 1852-53* (Manchester, T. Sowler & Sons, 1852), p. 73 and in *Calendar for the session 1853-54* (Manchester, T. Sowler & Sons, 1853), pp. 73-4.

35. Thompson, *Owens College*, p. 148.

36. Owens College, 'Examination papers', *Calendar for the session 1854-55* (Manchester, T. Sowler & Sons, 1854), pp. 91-3.

37. See H.E. Roscoe, *The life and experiences of Sir Henry Enfield Roscoe, DCL, LLD, FRS* (London, Macmillan and Co., 1906).

38. The appeal to local pride in Roscoe's biography of Dalton was quite unashamed. 'Whether in fact Dalton and Joule would have accomplished their life-work more fully had they been born to the intellectual purple of the ancient universities, and had to spend their time amidst the, to many minds, somewhat enervating influences of college life, instead of in the more robust and stimulating air of sturdy northern independence and intelligent northern activity. [sic]' The grammatical incompleteness of the sentence highlights the certainty of the answer to the implied question. H.E. Roscoe, *John Dalton and the rise of modern chemistry* (London, Cassell, 1895), p. 11.

39. Roscoe, *Life and experiences*, pp. 105-6. Of course the student later came to a sticky end as Roscoe piously recounts. The role of research as moral training in Roscoe's thought is similar to that of his American contemporary, Ira Remsen. Owen Hannaway, 'The German model of chemical education in America: Ira Remsen at Johns Hopkins', *Ambix*, **23** (1976), 145-64.

40. See Roscoe, *Life and experiences*. The three volumes of H.E. Roscoe and C. Schorlemmer, *A treatise on chemistry* (London, Macmillan and Co.), which first appeared in 1877-84 and went through many editions, were known to generations of students as 'Roscoe and Schorlemmer' even long after the latest edition had ceased to have any relation to the original authors. This meant that, well into the twentieth century, Roscoe's name was on a par with that of the great German encyclopaedists such as Gmelin.

41. For Roscoe's consulting activities, see Michael Sanderson, *The universities and British industry, 1850-1970* (London, Routledge and Kegan Paul, 1972), pp. 83-4.

42. Noted as an aside in Joseph Thompson's notes of the Owens College Trustees Minutes, MSS 378.42, M60/3. vol. 3, pp. 55-6; Archives, Manchester Central Reference Library.

43. On the condition of the College when Roscoe arrived, see R.H. Kargon, *Science in Victorian Manchester*, pp. 164-7.

44. 'Royal Commission on scientific instruction and the advancement of science', vol. 1, 'First, supplementary and second reports, with minutes of evidence and appendices', *P.P.* 1872 [536] XXV. 1, [Devonshire Commission], q. 7366, gives the annual enrolment in chemistry classes. Overall enrolment is given in J. Thompson, *Owens College*, p. 472.

45. 'Report of the Select Committee on Scientific Instruction', [Samuelson Committee], *P.P.* 1867-68 [432] XV. 1, qq. 5600-2. The proportion of occasional to full-time students can be deduced from J. Thompson, *Owens College*, pp. 472-3 and from the detailed figues given in the annual *Calendars* for the 1870s.

46. Owens College, *Calendar for the session 1862-63* (Manchester, T. Sowler & Sons, 1862), pp. 33-5.

47. *Ibid.*, p. 35.

48. H.E. Roscoe, 'Lecture Notes 1857-58', MS CH R106, John Rylands Library, University of Manchester.

49. This is clear from his autobiography and articles such as his 'Original research as a means of education' in *Essays and addresses by the staff of Owens College* (Manchester, 1874) and his *Record of work done in the Chemical Department of Owens College 1857-1887* (London, Macmillan and Co. for private circulation, 1887).

50. See for example H.E. Roscoe, *Record of work done*, pp. 9-10.

51. See H.E. Roscoe, *Record of work done;* 'Original research'; and 'Introductory lecture on the development of physical science: Delivered at Owens College Manchester', Johns Rylands Pamphlets 5.

52. See for example William Gregory, *Letter to the Rt. Hon. George, Earl of Aberdeen on the state of the schools of chemistry in the United Kingdom* (London, Taylor and Walton, 1842).

53. Otto Sonntag, 'Liebig on Francis Bacon and the utility of science', *Ann. Sci.*, **5** (1974), 373-86.

54. Devonshire Commission, vol. 1, q. 7390.

55. A.W. Williamson, *'A plea for pure science': Being the inaugural lecture at the opening of the Faculty of Science in University College, London* (London, Taylor and Francis, 1870). It is interesting that the same title was chosen by the American physicist, Henry Rowland, in his much more famous appeal a decade later; Henry A. Rowland, 'A plea for pure science' in *AAAS Proceedings, 1883* (Salem, Mass., AAAS, 1884).

56. For a list see *Chem. News,* **8** (26 April 1863), 153-5.

57. See M. Sanderson, *The universities,* chapters 2-4, passim; and W.H.G. Armytage, *Civic universities* (London, Routledge and Kegan Paul, 1965).

58. T.E. Thorpe, *The Right Honourable Sir Henry Enfield Roscoe: A biographical sketch* (New York, Longmans, Green & Co., 1916), p. 108.

59. For a description of the Owens College laboratories see H.E. Roscoe, 'Description of plans of the chemical laboratories at the Owens College, Manchester', n.d., Science Museum Library.

60. Sir T. Wemyss Reid, *Memoirs and correspondence of Lyon Playfair* (London, Cassell & Co, 1899).

61. Lyon Playfair, *A century of chemistry in the University of Edinburgh. Being the introductory lecture to the course of chemistry in 1858* (Edinburgh, Murray and Gibbs, 1863).

62. R.G.W. Anderson, *The Playfair Collection and the teaching of chemistry at Edinburgh, 1713-1858* (Edinburgh, The Royal Scottish Museum, 1978).

63. See *London Post Office Directory* (London, Kelly & Co, 1854), p. 1394. In that year only eleven analytical chemists were listed for London.

64. Chemical Society, 'Minutes of Council', 22 May 1848, p. 41, Royal Society of Chemistry Library, London.

65. Lyon Playfair, *Industrial instruction on the continent. Being the introductory lecture of the session 1852-53, Government School of Mines and of Science Applied to the Arts* (London, HMSO, 1854).

66. For an admission that he had promoted himself, see L. Playfair to A. Ramsay, n.d. January 1842, Ramsay MSS 11574, no. 2, National Library of Wales, Aberystwyth. Jack Morrell and Arnold Thackray, *Gentlemen of science: Early years of the British Association for the Advancement of Science* (Oxford, Clarendon Press, 1981), pp. 489-90.

67. T. Wemyss Reid, *Lyon Playfair*, p. 44.

68. Playfair's proposal is noted in L. Playfair to W. Buckland, 17 October 1842, Add. MSS 40517, f. 110, British Library. For evidence of Playfair's active politicking, see L. Playfair to J. Liebig, 15 November 1841 and 27 December 1841, Liebigiana 58, nos. 6 and 7, Deutsches Museum, Munich. See also L. Playfair to A. Ramsay, 1 July 1842, 13 October 1842, and 28 November 1842, Ramsay MSS 11574, nos. 22, 29, and 35.

69. T.Wemyss Reid, *Lyon Playfair*, pp. 91-108.

70. H. De la Beche and L. Playfair, 'First report on the coals suited to the steam navy ...' in *Mem. Geol. Survey G.B.*, **2**, pt. 2 (1848), 539-630.

71. L. Playfair to H. De la Beche, 'Annual report of the Chemical Department, 1849-50', n.d., GSM 1/6, pp. 34-7, Institute of Geological Sciences Library.

72. For the ambitions of Playfair and Joule see their first paper, 'On atomic volume and specific gravity', *Mem. Proc. Chem. Soc.*, **2** (1843-5), 401-3.

73 Cited in H. De la Beche to Lord Seymour, 20 May 1851, GSM 1/6, pp. 66-9, Institute of Geological Sciences Library. This could hardly be classed as a thriving school.

74. T.W. Phillips to H. De la Beche, 25 June 1851, *ibid.*, pp. 90-8.

75. L. Playfair to H. De la Beche, 10 February 1852, *ibid.*, pp. 132-4. See also the essay on Playfair in J.G. Crowther, *Statesmen of science* (London, Cresset Press, 1965).

76. 'Chemical examination', q. 8 in L. Playfair, *Industrial instruction on the continent*, Appendix.

77. L. Playfair to A. Ramsay, n.d. January 1842, Ramsay MSS 11574, no. 2.

78. On discussions following the Great Exhibition, see J.G. Crowther, *Statesmen of science*.

79. L. Playfair to H. De la Beche, 20 August 1851, printed in T. Wemyss Reid, *Lyon Playfair*, pp. 134-5.

80. Lyon Playfair, 'The study of abstract science essential to the progress of industry: Introductory lecture at the Government School of Mines, session 1851-52', in *British eloquence: Lectures and addresses* (London, 1855).

81. Lyon Playfair, 'The chemical principles involved in the manufactures of the Exhibition', in *Lectures on the results of the Exhibition delivered before the Society of Arts, Manufactures and Commerce at the suggestion of Prince Albert* (London, Society of Arts, 1852). For a judgement on the rhetorical status of Playfair's speech and a more detailed exposition of events after the Great Exhibition, D.S.L. Cardwell, *The organisation of science in England*, 2nd ed. (London, Heinemann, 1972), pp. 79-85.

82. Lyon Playfair, *Industrial instruction on the continent*.

83. On Playfair's vision of a central teachers' college see David Layton, *Science for the people: The origins of the school science curriculum in England* (London, George Allen and Unwin, 1973), pp. 144-66.

84. L. Playfair to H. Cole, 17 March 1853, Cole Papers, Correspondence, Box 15, Victoria and Albert Museum.

85. R. Murchison to R. Phillips, 18 April 1855 in A. Geikie, *Life of Roderick Murchison based on his journals and letters with notices of his scientific*

contemporaries and a sketch of the rise of palaeozoic geology in Britain (2 vols., London, John Murray, 1875), vol. 2, pp. 187-8.

86. For a fuller exposition of the Byzantine politics of this period see *ibid.*, pp. 184-92; and D. Layton, *Science for the people*, p. 155.

87. Board of Trade Department of Science and Art, 5 July 1853, Great Britain, Public Record Office: Ed. 28/1, f. 203.

88. D. Layton, *Science for the people*, pp. 129-63, discusses the failure of the science schools, particularly the complex political reasons underlying it.

89. The Watt school was the first to be awarded a science scholarship by the Science and Art Department during Playfair's tenure, 'First Report of the Department of Science and Art', *P.P.* 1854 [1783] XXVIII. 294, pp. 385-97.

90. Rachel E, Waterhouse, *The Birmingham and Midland Institute* (Birmingham, Birmingham and Midland Institute, 1954), p. 29.

91. Lyon Playfair, *Science in its relations to labour. Being a speech delivered at the anniversary of the People's College Sheffield, 25 October 1853* (London, Chapman and Hall, 1853), p. 11. On the College see G.C. Moore Smith, *The story of the People's College, Sheffield, 1842-1878* (Sheffield, J.W. Northend, 1912).

92. Devonshire Commision, vol. 1, qq. 8509-12.

93. 'Fifth report of the Department of Science and Art', *P.P.* 1857-58 [2385] XXIV. 219, p. 22.

94. L. Playfair to Professor Keeland, 3 August 1858 in University of Edinburgh, Senate Minutes, **1** (1855-61), pp. 225-6; see also 27 October 1858; for the discussion and for formal approval see *ibid.*, 6 November 1858, p. 235. University of Edinburgh Archives, University of Edinburgh Library.

95. 'Royal Commission on scientific instruction and the advancement of science. Minutes of evidence, appendices and analyses of evidence,' vol. 2 [Devonshire Commission, Evidence 2], *P.P.* 1874 [c. 958] XXII. 1, q. 9345.

96. Playfair used the standard Department of Science and Art syllabus and examinations, L. Playfair to H. Cole, 10 March 1859, Cole Papers, Box 15. He described his teaching to the Samuelson Committee, qq. 1032-43.

97. Playfair had also been the promoter of a committee to investigate the possibility of science degrees and was convener of that committee when it was founded, University of Edinburgh, Senate Minutes, **2** (1861-5), 31 January 1863, p. 246; and 28 February 1863, p. 261. See also *Edinburgh University Calendar for the year 1865-66* (Edinburgh, Maclachlan and Stewart, 1865).

98. Money was sought from the Highland and Agricultural Society for an increased endowment of the Chair of Agriculture (Professors Christison, Playfair and Wilson were sent as envoys), University of Edinburgh, Senate Minutes, 30 November 1867, p. 330. Discussions with the linen tycoon David Baxter (*DNB*) over support for a Chair of Engineering were also led by Playfair and Wilson, University of Edinburgh, Senate Minutes, **3** (1865-9), 20 January 1868, p. 390. See also J.B. Morrell, 'The patronage of mid-Victorian science in the University of Edinburgh' in G.L.'E. Turner, ed., *The patronage of science in the nineteenth century*, Science in History 1 (Leyden, Noordhof International Publishing, 1976), pp. 53-93.

99. University of Edinburgh, Senate Minutes, **3** (1865-9), 8 February 1868, p. 413, for the reading of the offers. On Playfair and Wilson drafting the

memorial to the Treasury see *ibid.*, 21 December 1867, p. 371. The memorial is recorded in *ibid.*, 7 January 1868, pp. 391-3. For Brewster's own interests in utility see his review of Whewell's *History of the inductive sciences* in *Edin. Rev.*, **66** (1837-8), 110-51.

100. University of Edinburgh, Minute Book of the University Court, **1** (1859-70), 24 April 1868, pp. 359-60. University of Edinburgh Archives.

101. *Edinburgh University Calendar for the year 1868-9* (Edinburgh, Maclachlan and Stewart, 1868), p. 26. For the claim that degrees in applied science were unconstitutional see University of Edinburgh, Senate Minutes, 26 October 1868, p. 499.

102. Samuelson Committee, q. 1042.

103. *Ibid.*, q. 1036.

104. For Playfair's proposal see above, n. 68.

105. H. Watts to H. E. Roscoe, 21 December 1864, Roscoe papers, Royal Society of Chemistry.

106. H.E. Roscoe, 'Lyon Playfair', *Nature*, **58** (1898), pp. 128-9.

107. R. Sviedrys, 'The rise of physical science at Victorian Cambridge' *Hist. Stud. Phys. Sci.* **2** (1970), pp. 127-45; *idem*, 'The rise of physics laboratories in Britain', *ibid.*, **7** (1976), 405-36.

108. Science and Art Department of the Comittee of Council on Education, 'Aid to scientific instruction', Science Form no. 14, July 1859, GSM 15, f. 97, Institute of Geological Sciences.

109. P.S. Uzzell, 'The Science and Art Department and the teaching of chemistry', *Vocational Aspect*, **29** (1977), 127-32. Uzzell suggests that the level was between the levels of GCE O and A level examinations.

110. C.J. Woodward, *Evening chemistry teaching in Birmingham, a forty years retrospect: An address to the Municipal Technical School Chemistry Society, 23 January 1900* (Birmingham, White and Pike, 1900), p. 12.

111. For Hofmann's opinion see 'Twelfth report of the Science and Art Department of the Committee of Council on Education', *P.P.* 1865 [3476] XVI. 301, App. B, p. 65.

112. For Frankland's opinion, see 'Sixteenth report of the Science and Art Department of the Committe of Council on Education', *P.P.* 1868-69 [4136] XXIII. 131, App. B, p. 86.

113. Figures on the numbers of students studying the various subjects are compiled from the annual reports of the Science and Art Department.

Chapter four

1. 'Tenth report of the Science and Art Department of the Committee of Council on Education', *P.P.* 1863 [3143] XVI. 21, App. B, 46.

2. 'Report of the Royal Commission on scientific instruction and the advancement of science', vol. 1, 'First, supplementary, and second reports, with minutes of evidence and appendices' [Devonshire Commission], *P.P.* 1872 [536] XXV. 1, q. 8995.

3. Medallists' careers are given in the annual reports of the Science and Art Department.

4. Frankland discusses his pupils in 'Report from the Select Committee on scientific instruction; with the proceedings of the committee, minutes of evidence and appendix' [Samuelson Committee], *P.P.* 1867-8 (432 and 432-I) XV. 1, q. 8049 and App. 18. A larger sample is considered in Devonshire Commission, vol. 1, q. 5682.

5. Devonshire Commission, vol. 1, q. 7365.

6. The composition of the student body is recorded in the annual reports of the Science and Art Department. We have compiled our list for 1845-1853 and 1853-1870 from Royal College of Chemistry, Registers of Students, College Archives, Imperial College London.

7. The total time attended by students at Owens can be calculated from Devonshire Commission, vol. 1, q. 7365, and from J. Thompson, *The Owens College: Its foundation and growth and its connection with the Victoria University Manchester* (Manchester, J.E. Cornish, 1886), pp. 153 and 172. A breakdown is also given in the annual Owens College *Calendars*.

8. Devonshire Commission, vol. 1, q. 5682; Samuelson Committee q. 8049 and App. 18.

9. G.W. Roderick and M.D. Stephens, *Education and industry in the nineteenth century: The English disease?* (London, Longman, 1978), p. 71.

10. Devonshire Commission, qq. 9162-3.

11. *Sir Henry Bessemer, FRS: An autobiography* (London, Offices of *Engineering*, 1905). Royal College of Chemistry, Registers lists Bessemer as an occasional student during four terms in 1857 and 1858, College Archives, Imperial College London.

12. John Morrison, 'On the manufacture of caustic soda', *Chem. News*, **31** (26 February 1875), p. 87.

13. Devonshire Commission, qq. 5511, and 5509.

14. [The Chemical Society], *Jubilee of the Chemical Society of London. Records of the proceedings together with an account of the history and development of the Society, 1841-1891* (London, Chemical Society, 1896), pp. 186-7.

15. Chemical Society application forms, 1841-1870, Royal Society of Chemistry.

16. W.A. Miller, 'Report of the President and Council', *Quart. J. Chem. Soc.*, **10** (1858), p. 187.

17. C. Daubeny, 'Report of the President and Council', *Quart. J. Chem. Soc.*, **6** (1854), p. 165.

18. R.F. Bud, 'The Royal Manchester Institution' in D.S.L. Cardwell, ed., *Artisan to graduate: Essays to commemorate the foundation of the Manchester Mechanics' Institution, now in 1974 the University of Manchester Institute of Science and Technology* (Manchester, Manchester University Press, 1974), pp. 119-33. See also Robert H. Kargon, *Science in Victorian Manchester: Enterprise and expertise* (Manchester, Manchester University Press, 1977).

19. R.H. Kargon, *Science in Victorian Manchester*, pp. 141-3.

20. A. Gibson and W.V. Farrar, 'Robert Angus Smith and sanitary science', *Notes Recs. Roy. Soc.*, **28**, pt.2 (1974), 241-62.

21. D.W.F. Hardie, *A history of the chemical industry in Widnes* (Widnes, ICI General Chemicals Division, 1950), p. 227.

22. R.H. Kargon, *Science in Victorian Manchester*, p. 149.

23. *J. Chem. Soc.*, **20** (1867), 392.

24. *J. Chem. Soc.*, **47** (1885), 331.

25. W. Crookes to P. Spence, 31 October 1864 in E.E. Fournier D'Albe, *The life of Sir William Crookes* (London, T. Fisher Unwin, 1923), pp. 86-8.

26. James Adams, *Biographical sketch of the late Frederick Penny, PhD, FRSE* (Glasgow, privately printed, 1870).

27. For Muspratt's school see G.W. Roderick and M.D. Stephens, *Scientific and technical education in nineteenth century England: A symposium* (New York, Barnes and Noble, 1977), pp. 65-73.

28. Alex E. Tucker, *Our Society, a lecture to the Birmingham Midland Institute*, 12 February 1902.

29. *Chem. News*, **45** (1882), 108.

30. E.E. Fournier D'Albe, *Crookes*, pp. 31-2 discusses science teaching at Chester.

31. For Gatebeck see J.D. Marshall and M. Davies-McShiel, *Industrial archaeology of the Lake counties* (Newton Abbot, David and Charles, 1969), p. 76. See also Paul N. Wilson, 'Gunpowder mills of Westmorland and Furness', *Trans. Newcomen Soc.*, **36** (1963-4), 47-69. For complaints about the high price of nitrate see 'Copy of a letter received from Messrs. James Buchanan & Co.', 8 August 1859, Melfort Gunpowder Co. Letterbook, 1859-60; Charcoal Iron Co. Records, BPB/2, p. 127; Cumbria Record Office, Barrow. A.D. Keightley, one of the Chemical Society members, gave as an address 'Old Hall', Kendal, which was the address of Edward Johnson, a Gatebeck partner. See Belle Halliwell, 'Story of Sedgwick Gunpowder Works', *Cumbria*, **123** (April 1962), 24-5.

32. For membership of the Whalley Club, see R.S. Crossley, *Accrington captains of industry* (Accrington, privately pub., 1930), p. 39 Another cluster of members was drawn from the Liverpool area.

33. On the scientific complexity of dye making even before synthetic dyestuffs, see C.M. Mellor and D.S.L. Cardwell, 'Dyes and dyeing, 1775-1860', *Brit. J. Hist. Sci.*, **1** (1963), 265-79. Geoffrey Turnbull, *A history of the calico printing industry of Great Britain* (Altrincham, Sherratt and Son, 1951) highlights the variety of problems and innovations that characterised the calico printing industry.

34. On the gift by Neild see University College London, *Proceedings of the annual general meeting* (London, Richard Taylor and Co, 1847), p. 7. On the gift by Thomson see Royal College of Chemistry, General Meeting, 31 August 1846, 'Minutes of the Council of the College', Imperial College London, College Archives.

35. Dale notebooks recently acquired by the North-West Museum of Science and Industry indicate the research carried out at Dale's works. It was there that Caro discovered what would be called 'Manchester Brown'.

36. Guy Christie, *Storey's of Lancaster, 1848-1964* (London, Collins, 1964), 48-50.

37. The schools represented by teachers in the Chemical Society by 1870 included: Wellington College, Worthing College, Hurstpierpoint, London International College, Queenwood College, Brighton College, University College School, Priory School, Westminster College, Stonyhurst College,

Chester College, King Edward's School (Birmingham), Halifax School, Huddersfield College, Wesleyan College (Taunton).

38. W. Crookes to R.A. Smith, 31 October 1864, in E. E. Fournier D'Albe, *Crookes*, p. 89.

39. For the development of structural theory in London during the 1850s see C.A. Russell, *The history of valency* (Leicester, Leicester University Press, 1971); and Alan J. Rocke, 'Origins of the structural theory in organic chemistry', PhD. Dissertation, University of Wisconsin-Madison, 1975.

40. For Kekulé's stay in London see Richard Anschütz, *August Kekulé* (2 vols. Berlin, Verlag Chemie, 1929), vol. 2, pp. 37-53.

41. D.S.L. Cardwell, *The organisation of science in England*, 2nd ed. (London, Heinemann, 1972), p. 124.

42. F. Knapp, *Lehrbuch der chemischen Technologie* (2 vols. Brunswick, 1847-48). F. Knapp, *Chemical technology or chemistry applied to the arts and manufactures* ed. and trans. Edmund Ronalds and Thomas Richardson (3 vols., London, Hippolyte Baillère, 1848-51). Knapp's book was part of a well-recognised genre of writing on the 'natural history of industry'.

43. For an explanation of the place of the alkali industry within the chemical industry see the work of the great Georg Lunge, who was a chemical manufacturer in South Shields before becoming Professor of Chemistry at the ETH in Zurich, *A theoretical and practical treatise on the manufacture of sulphuric acid and alkali* (3 vols., London, John Van Voorst, 1879-80), vol. 1, pp. 1-5.

44. Statistics on the industry are from L.F. Haber, *The chemical industry during the nineteenth century* (London, Oxford University Press, 1958), pp. 18, 59; and E. Schunk, R.A. Smith and H.E. Roscoe, 'On the recent progress and present condition of manufacturing chemistry in the South Lancashire district', BAAS, *Reports* (1861), pp. 108-28 (115).

45. L .F. Haber, *Chemical industry*, p. 104.

46. See for instance G.L.M. Strauss, *England's workshops* (London, Groombridge and Sons, 1864), pp. 133-84. In this not untypical journalistic report on several industries there is a revealing comment at the end of the treatment of the chemical industries. 'We have said nothing about the many admirable mechanical appliances used in Messrs. Huskisson's manufactory, feeling that the chemical portion of the subject would be of more interest to our readers' (p. 169).

47. See G. Lunge, *Manufacture of sulphuric acid and alkali, passim.*

48. There are relatively few useful analyses of the patent literature. There is, however, the classic S.C. Gilfillan, *The sociology of invention* (Cambridge, Mass., MIT Press, 1935), reprinted 1963. For more recent work see H. Van den Belt, E. Homburg, and W.J. Hornix, 'The development of the dye industry', Progress Report, Faculty of Science, University of Nijmegen, December 1981; and F. Leprieur and P. Papon, 'Synthetic dyestuffs: The relations between academic chemistry and the chemical industry in nineteenth-century France', *Minerva*, **17** (1979), 197-224.

49. Helpful background is in Neil Davenport, *The United Kingdom patent system, a brief history with bibliography* (Havant, K. Mason, 1979). A useful

guide to the intricacies of patent law is R. Ellis, *The patent law in relation to chemistry* (London, Gill and Ellis, 1912).

50. 'Report of the Commission appointed to inquire into the working of the law relating to Letters Patent for Invention', *P.P.* 1864 [3419] XXIX. 321, q. 529.

51. *Abridgements of specifications relating to acids, alkalies, oxides and salts* [Class 40] (4 vols. London, Office of the Commissioners of Patents for Invention, Part 1, 1622-1866; Part 2, 1867-76). See Appendix D for details of our analysis.

52. H. van den Belt, E. Homburg, and W.J. Hornix, 'Dye industry'.

53. Frank Spence, 'Manchester Society for Promoting Scientific Industry', *Manchester Guardian*, 16 October 1872. One of Spence's arguments for the Society was the 'threat' posed by the relative rates of growth of numbers of American and British patents. The number of patents issued by foreign countries was also noted with concern by the Commission on Patent Law, 1864, p. 77.

54. A.E. Tucker, 'Our Society'.

55. Black Ash, 'Proposed Association of Manufacturing Chemists', *Chem. News*, **26** (1872), p. 46. This move led to the formation of the Society for the Promotion of Scientific Industry; W.V. Farrar, 'Society for the Promotion of Scientific Industry, 1872- 1876', *Ann. Sci.*, **29** (1972), 81-6.

56. The earliest known use of the term 'chemical engineer' seems to have been in the article on sulphuric acid in A. Ure, *Dictionary of arts, manufactures, and mines* (London, 1839), p. 1220, cited in a letter to the editor by S.A. Gregory, 'Chemical Engineers', *Chem. & Ind.*, **6** (October 1951), 838.

57. George E. Davis, *A handbook of chemical engineering* (2 vols., Manchester, Davis Bros., 1901), vol. 1, p. 4.

58. This recognition of Davis is reflected throughout the articles in the recent American Chemical Society symposium, William Furter, ed., *History of chemical engineering*, ACS Advances in Chemistry Series, no. 190 (Washington, D.C., ACS, 1980). See especially, F.J. Van Antwerpen, 'The origins of chemical engineering', pp. 1-14; John T. Davies, 'Chemical engineering: How did it begin and develop?', pp. 15-44; and D.C. Freshwater, 'George E. Davis, Norman Swindin and the empirical tradition in chemical engineering', pp. 97-112.

59. For the life of Davis see D.W.F. Hardie, 'The empirical tradition in chemical technology', first Davis-Swindin Memorial Lecture, Loughborough College of Technology, 1962.

60. George E. Davis, *Handbook*, vol. 1, p. 4.

61. D.W.F. Hardie, 'Empirical tradition', pp. 9-10.

62. D.W.F. Hardie and J. D. Pratt, *A history of the modern British chemical industry* (London, Pergamon Press, 1966), p. 95 make the point well. 'It would not be too much to say that, until almost the end of the last century, chemical industry, particularly in its heavy branch, was largely engineering: chemical plants were viewed as comparatively simple machines or structures, furnaces and pots in which chemical changes were brought about. Practically all the advances made in heavy chemical manufacture were improvements of

means, mechanical and structural: the engineers, in the old chemical industry, had something like the impact of Ransome, Croskill and Tull on the advance of agriculture.'

63. E. Schunck, R.A. Smith and H.E. Roscoe, 'Manufacturing chemistry in South Lancashire'.

64. P.J. Worsley to A. Chance, 7 November 1876, DIC/BM20/186, Cheshire Record Office. On the innovative record of the Chance company see L.F. Haber, *Chemical industry*, pp. 161-2.

65. A.M. Chance, 'Manufacture of caustic soda', 15 November 1876, DIC/BM20/ 186, Cheshire Record Office.

66. H. Brunner to L. Mond, 3 June 1876, DIC/BM7/1, Cheshire Record Office. See also a similar letter in the same file, 14 June 1876.

67. For the price in 1886 see 'Report of John Brunner', 14 October 1887, DIC/BM20/186, Cheshire Record Office. This rapid decline in prices is discussed in wider context by L.F. Haber, *Chemical industry*, pp. 211-13.

68. 'Exhibition of appliances for the economical consumption of fuel at Manchester', *The Engineer*, **37** (1874), 2.

69. Frank, 'Manufacture of sulphate of ammonia - proposed Association of Manufacturing Chemists', *Chem. News*, **25** (1872), 309.

70. The outstanding source for the history of the alum industry is Charles Singer, *The earliest chemical industry: An essay in the historical relations of economics and technology illustrated from the alum trade* (London, The Folio Society, 1948); for the extraction of ammonia from gas works waste, pp. 283-8; for Frank Spence's ammonia process, p. 288. The best detailed treatment of Spence's works is T. Lenten Elliot, 'Alum manufacture during the last hundred years', typescript in Peter Spence Archives. This and the research notebooks are preserved at the company in Widnes, now a subsidiary of Laporte PLC. We are most grateful to Dr. Allan Comyns for giving us access and assistance. See also J. Fenwick Allen, *Some founders of the chemical industry*, 2nd ed., (London, Sherratt and Hughes, 1907), pp. 251-89.

71. 'Index of patents', Chance and Hunt, December 1885, DIC/BM20/219, Cheshire Record Office.

72. D.W.F. Hardie and J.D. Pratt, *History of British chemical industry*, p. 52.

73. There is no satisfactory account of this works which was for almost a hundred years the largest gas works in the world, but see the talk given to the 95th General Meeting of the Institution of Gas Engineers by L.W. Blundell, *By-products in the economy of gas manufacture*, copyright publication no. 528 (London, Institution of Gas Engineers, 1958). See also the illustrated booklet *Beckton 1868-1968* (London, North Thames Gas Board). On tar production see L.F. Haber, *Chemical industry*, pp. 163-4, and D.W.F. Hardie and J.D. Pratt, *History of British chemical industry*, p. 52.

74. A.G. Bloxam, 'Patent law in relation to the dyeing industry' in W.M. Gardner, *The British coal-tar industry, its origin, development, and decline* (London, Williams and Norgate, 1915), pp. 169-79. For our figures which give a similar pattern, see Appendix D.

75. Dale Notebooks, North West Museum of Science and Industry.

76. Samuelson Committee, qq. 5509-10.

77. For German production of dyestuffs, see L.F. Haber, *Chemical industry*, p. 128, Table 3; for British soda production, p. 213, Table 3 and p. 214, Table 4. In 1890 each industry had sales of about £3 million.

78. Samuelson Committee, q. 5511.

79. The publications of Chemical Society manufacturers listed in the Royal Society *Catalogue* tend to be, where chemical, contributions on specific (often analytical) problems arising within a particular applied field.

80. R.C. Chirnside and J.H. Hamence, *The 'Practising Chemists': A history of the Society for Analytical Chemistry, 1874-1974* (London, SAC, 1974), pp. 83-6.

81. On specialisation during this period see Aaron J. Ihde, *The development of modern chemistry* (New York, Harper and Row, 1964), Part III. In recent years, there has been considerable scholarly preoccupation with the history of various of the specialties. See for example R.G.A. Dolby, 'Debates over the theory of solution: A study of dissent in physical chemistry in the English-speaking world in the late nineteenth and early twentieth centuries', *Hist. Stud. Phys. Sci.*, 7 (1976), 297-404; Robert Kohler, *From medical chemistry to biochemistry: The making of a biomedical discipline* (Cambridge, Cambridge University Press, 1982).

82. A contemporary appreciation of the process of specialisation is that of Roscoe, who once referred in conversation to the Professor of Organic Chemistry at Owens College as '... my Master, he is much better in Organic Chemistry than I ...', note by J. Thompson in Minute Book of the Proceedings of Council, Owens College, no.1, 12 June 1874, Manchester Central Library, MSS 378. 42, M60/7.

83. M.C. King, 'Experiments with time', *Ambix*, 28 (1981), 70-82; *idem*, 'Progress and problems in the development of chemical kinetics', *Ambix*, 29 (1982), 49-61.

Chapter five

1. On Oxford, see Harold Hartley, 'Schools of Chemistry of Great Britain and Ireland - XVI: The University of Oxford', *J. Roy. Inst. Chem.*, 79 (1955), 116-27, 176-84. On Cambridge, see W.H. Mills, 'Schools of Chemistry in Great Britain and Ireland - VI: The University of Cambridge', *ibid.*, 77 (1953), 423-31, 467-73; and G.K. Roberts, 'The liberally educated chemist: Chemistry in the Cambridge Natural Sciences Tripos, 1851-1914', *Hist. Stud. Phys. Sci.*, 11 (1980), 157-83.

2. 'Report of the Commissioners appointed to inquire into the revenues and management of certain schools and the studies pursued and instruction given therein' [Clarendon Commission], vol. 4, 'Evidence, part 2', *P.P.* 1864 [3288] XXI.537, qq. 27-31.

3. F.W. Farrar, *Essays on a liberal education* (London, Macmillan, 1867).

4. Edward C. Mack, *The Public Schools and British opinion since 1860: The relationship between contemporary ideas and the evolution of an English institution*, (New York, Columbia University Press, 1941), pp. 50-90. Brian Simon, *Studies in the history of education, 1780-1870* (London, Lawrence & Wishart, 1960), chapter 6.

5. 'Fifth Report of the Science and Art Department of the Committee of Council on Education', *P.P.* 1857-58 [2385] XXIV. 219, p. 22.

6. *Report of the Committee appointed by the Council of the Society of Arts to inquire into the subject of industrial instruction with the evidence on which the report is founded* (London, Longman, Brown, Green and Longman for the Council of the Society of Arts, 1853), p. 8.

7. 'What is true technical education?', *The Economist*, **26** (25 January 1868), 87-8.

8. C. Critchett, ed., *Reports of artizans selected by a committee appointed by the Council of the Society of Arts to visit the Paris Universal Exhibition, 1867* (London, Society of Arts, 1867). See also the little known but still useful D.H. Thomas, 'The development of technical education in England from 1851-1889 with special reference to economic factors', PhD Dissertation, London School of Economics, 1940; and S. Cotgrove, *Technical education and social change* (London, Allen and Unwin, 1958).

9. 'Thirteenth Report of the Science and Art Department of the Committee of Council on Education', *P.P.* 1866 [3649] XXV. 337, App. B, pp. 394-5.

10. 'Advertisement', *The Chemist*, **5** (1844), p. 2.

11. J.F. Donnelly, 'Memorandum of suggestions for enlarging the system of state aid to scientific instruction, drawn up in accordance with the instruction of the Lords of the Committee of Council on Education', in 'Report from the Select Committee on scientific instruction for the industrial classes together with the proceedings of the committee, minutes of evidence and appendix' [Samuelson Committee], *P.P.* 1867-68 (432 and 432-I) XV. 1, App. 11, p. 449.

12. See P.W. Musgrave, 'The definition of technical education, 1860-1910', *Vocational Aspect*, **16** (1964), 105-11; reprinted in P.W. Musgrave, ed., *Sociology, history and education: A reader* (London, Methuen, 1970), pp. 65-74.

13. 'Seventh Report of the Science and Art Department of the Committee of Council on Education', *P.P.* 1860 [2626] XXIV. 7, App. B, p. 108.

14. 'Memorial. To the Rt. Hon. Earl Granville, K.G., the President of the Council on Education, and the Rt. Hon. W.F. Cowper, M.P., the Vice-President of the Council on Education, the memorial of working men belonging to the Royal Polytechnic, the London Mechanics', and other Institutions in London', reprinted in Samuelson Committee, App. 19, p. 474.

15. On the effects of the Revised Code, see David Layton, *Science for the people* (London, George Allen and Unwin, 1973); and P.W. Musgrave, *Society and education in England since 1800* (London, Methuen, 1968), pp. 36-7. For the effect on teachers, see A. Tropp, 'The changing status of the teacher in England and Wales', in P.W. Musgrave, ed., *Sociology, history and education*, pp. 198-202. For contemporary views, see for example Samuelson Committee, p. iv and q. 8271.

16. Henry Cole and J.F. Donnelly, Samuelson Committee, qq. 79- 80, 663-4. See the 'Reports of the Commissioners appointed to make inquiry with regard to scientific instruction and the advancement of science, and to inquire what aid there is derived from grants voted by Parliament, or from

endowments belonging to the several Universities of Great Britain and Ireland and the colleges thereof, and whether such aid could be rendered in a manner more effectual for the purpose' [Devonshire Commission], vol. 1. 'First, supplementary, and second, reports, with minutes of evidence, appendices, and analyses of evidence', *P.P.* 1872 [536] XXV. 1, qq. 5965-6.

17. John Hubbel Weiss, *The making of technological man: The social origins of French engineering education* (Cambridge, Mass., MIT Press, 1982).

18. See J.F. Donnelly, Samuelson Committee, qq. 7868; and J.F. Donnelly, 'Memorandum on the Royal School of Mines, and Science School and Science Museum, South Kensington', in 'Correspondence between the Science and Art Department and the Treasury as to the organization of the Normal School of Science and Royal School of Mines', *P.P.* 1881 [c. 3085] LXXIII. 563, pp. 7-31 (9). A proof copy of the report, July 1861, can be consulted in Miscellanies, 13, f. 7, Cole Papers, Victoria and Albert Museum.

19. See also Geoffrey Dyson, 'The development of instruction in naval architecture in 19th century England', MA Thesis, University of Kent at Canterbury, 1978, chapter 4.

20. George S. Emmerson, *John Scott Russell: A great Victorian engineer and naval architect* (London, John Murray, 1977), pp. 180-1.

21. 'Twelfth Report of the Department of Science and Art of the Committee of Council on Education', *P.P.* 1865 [3476] XVI. 301, App. B.

22. B.B. Kelham, 'Science education in Scotland and Ireland', PhD Dissertation, University of Manchester, 1968.

23. F. Peel to Committee of Council on Education, 13 May 1862; Treasury Out-Letters to the Privy Council, Great Britain, Public Record Office: T.9/11 (1857-63), f. 425. See also Treasury Minutes, T.29/588, p. 43 which makes the origin of the Commission obvious, with references to Donnelly's involvement dated 11 December 1861, 26 February 1862 and 10 April 1862. Donnelly's report is included in Treasury Board Papers, In-Letters, T.1/ 6397B/18415.

24. The Treasury report is in T.1/6397B/18415. For the objections to it, see G.A. Hamilton to Committee of Council on Education, 23 April 1863, T.9/11, ff. 489-91. The parliamentary discussion is in 3 Hansard 174, 8 April 1864, cols. 663-74.

25. The brief is recorded in Miscellanies, 13, f. 179, Cole Papers, Victoria and Albert Museum. With slightly different wording it is repeated in 'Report on the College of Science for Ireland' *P.P.* 1867 (219) LV. 777, p. 1.

26. *Ibid.*, pp. 2-4.

27. D.S.L. Cardwell, *Organisation of science in England* 2nd ed. (London, Heinemann, 1972), pp. 111-12.

28. 'Industrial progress and the education of the industrial classes in France, Switzerland, Germany &c', *P.P.* 1867-68 (13) LIV. 67.

29. Anthony Bishop and Wilfred James, 'The Act that never was: The Conservative Education Bill of 1868', *Hist. Ed.*, 1 (1972), 160-73; Henry Roper, 'Towards an Elementary Education Act for England and Wales, 1865-1868', *Brit. J. Ed. Studies*, 23 (1975), 181-208; Paul Smith, *Disraelian Conservatism and social reform* (London, Routledge and Kegan Paul, 1967).

30. [Benjamin Disraeli], *The Chancellor of the Exchequer in Scotland, being*

the text of two lectures delivered by him in the City of Edinburgh on 29th and 30th October 1867 (Edinburgh, William Blackwood and Sons, 1867).

31. J. F. Donnelly, 'Memorandum on State Aid to Scientific Instruction', 12 November 1867, Samuelson Committee, App. 11, pp. 447-9; H. Cole, 'Notes on Public Education', 28 November 1867, *ibid.*, App. 12, pp. 459-61.

32. 'Copy of minute of the Lords of the Committee of Council on Education relating to scientific instruction and explanatory memorandum thereon', *P.P.* 1867-68 (193) LIV. 17.

33. 'Bill to regulate the distribution of sums granted by Parliament for elementary education in England and Wales and for other purposes', House of Lords Papers, 1867-68 (LIII). 227. The bill was published on 28 March 1868. On contemporary evaluations of Montagu, see P. Smith, *Disraelian Conservatism*, 69, 83; and John Vincent, ed., *Disraeli, Derby and the Conservative Party: Journals and memoirs of Edward Henry, Lord Stanley, 1849-1869* (Brighton, Harvester, 1978), p. 327. An entry for 18 January 1868 reads '... his [Montagu's] total want of tact makes it impossible that he should be entrusted with the conduct of an important measure'.

34. 'Answers from Chambers of Commerce to queries of the Vice-president of Council as to technical education', *P.P.* 1867-68 (168) LIV. 23.

35. See *J. Soc. Arts*, **16** (1868), 184-209 and 627-42; D.S.L. Cardwell, *The organisation of science*, pp. 111-15.

36. Minutes of the Owens College Extension Committee, 13 January 1868, vol. 1, pp. 37-40. The Committee had been encouraged to seek government aid by the funding of the Royal College of Science, Dublin; *ibid.*, 1 May 1867, vol. 1, p. 29. Manchester Central Reference Library Archives, MSS 378.42, M60/5.

37. *The Economist*, **26** (25 January 1868), 87-8.

38. Alfred Neild, 'Technical education', *The Economist*, **26** (8 February 1868), p. 153.

39. 3 Hansard 191, 24 March 1868, cols. 159-66.

40. *Ibid.*, cols. 170-9.

41. Roy MacLeod, 'Scientific and technical education' in G. Sutherland, ed., *Education, government and society in Britain: Commentaries on British Parliamentary Papers* (Dublin, Irish University Press, 1977), pp. 195-233 (201-2).

42. Henry Cole, Diaries, 16 May 1868, Victoria and Albert Museum .

43. Samuelson Committee, Cole qq. 79-80, 105 and 887; Frankland q. 8109; Playfair q. 1146.

44. *Ibid.*, qq. 293-300; 898-906.

45. *Ibid.*, q. 303.

46. *Ibid.*, q. 416.

47. *Ibid.*, q. 672.

48. *Ibid.*, qq. 432-9.

49. *Ibid.*, Playfair q. 1149; Frankland q. 8085.

50. *Ibid.*, qq. 1026-8.

51. *Ibid.*, q. 8113.

52. *Ibid.*, q. 5541.

53. *Ibid.*, p. viii.

54. *Ibid.*, p. vii.

55. *Ibid.*, p. ix.

56. *J. Soc. Arts*, **16** (1868), 627, 632.

57. *Ibid.*, p. 633.

58. H. Cole and J.F. Donnelly, 'Memorandum on the expectations which the late Government held out in 1868 in respect of aiding scientific instruction of an advanced character' 18 February 1869, Miscellanies 16, f. 18, Cole Papers, Victoria and Albert Museum. 'History of the Science and Art Department of the Committee of Council on Education since its creation', in 'Thirtieth Report of the Science and Art Department of the Committee of Council on Education', *P.P.* 1883 [c. 3618] XXVII. 1, p. xxxv. The grant was not taken up very rapidly, 'Sixteenth Report of the Science and Art Department of the Committee of Council on Education', *P.P.* 1868-69. XXIII. 131, App. B, p. 58.

59. *The Economist*, 24 October 1868, 1218-19. See also R. H. Kargon, *Science in Victorian Manchester: Enterprise and expertise* (Manchester, Manchester University Press, 1977), pp. 190-6.

60. 3 Hansard 198, 19 July 1869, cols. 161-2.

61. *Ibid.*, col. 218.

62. 'Report from the Select Committee on the South Kensington Museum, with the proceedings, minutes of evidence, appendix and index', *P.P.* 1860 (504) XVI. 527. The decision to build is considered in 'Correspondence between the Treasury and the Science and Art Department relative to new buildings for the South Kensington Museum', *P.P.* 1866 (432) XL. 441.

63. 'Thirteenth Report of the Science and Art Department of the Committee of Council on Education', *P.P.* 1866 [3649] XXV. 337, App. A-1, p. 3. H.C. Childers to Mr. Bruce, 2 May 1866 and 24 May 1866, Great Britain, Public Record Office: T.9/12, ff. 195-6, and f. 203 respectively. The story of the South Kensington complex is chronicled in Harry Butterworth, 'The Science and Art Department, 1895-1900', PhD Dissertation, Oxford University, 1969.

64. 'Fourteenth Report of the Science and Art Department of the Committee of Council on Education', *P.P.* 1867 [3853] XIII.1, p. xii.

65. 'Fifteenth Report of the Science and Art Department of the Committee of Council on Education', *P.P.* 1867-68 [4049] XXVII. 419, p. xii.

66. The details are recounted in 'Copy of any letters between the Science and Art Department and the Treasury on the subject of the estimate, plans, and models of the future buildings at South Kensington', *P.P.* 1870 (218) LIV. 737.

67. Samuelson Comittee, qq. 7861-77; Cole's memorandum was recorded in q. 7860. For the most part, Montagu did the talking, with Donnelly replying in monosyllables.

68. [Henry Cole], 27 March 1868, Miscellanies, 14, 438.

69. Henry Cole, Diaries, 16 May 1868.

70. 3 Hansard 199, 7 March 1870, col. 1363.

71. Henry Cole, Diaries, 5 May 1870.

72. D.S.L. Cardwell, *Organisation of science*, pp. 119-26; R. MacLeod, 'Scientific and technical education', pp. 203-10. See also A.J. Meadows,

Science and controversy: A biography of Sir Norman Lockyer (London, Macmillan, 1972), pp. 75-95.

73. Devonshire Commission, vol. 1, qq. 1239-59.

74. J.G. Crowther, *Statesmen of science* (London, Cresset Press, 1965).

75. The reports of the Devonshire Commission not yet cited can be located as follows: Third, *P.P.* 1873 [c. 868]. XXVIII; Fourth, fifth, and minutes of evidence, *P.P.* 1874 [c. 884], [c. 1087], [c. 958] XXII; Sixth, Seventh, Eighth and minutes of evidence, *P.P.* 1875 [c. 1279], [c. 1297], [c. 1298], [c. 1363]. XXVIII.

76. The Treasury's view of Cole at this time is clear from its letter books, see for example Great Britain, Public Record Office: T.9/13, 271 (23 December 1869), 345 (29 March 1870), 466-8 (18 November 1870), 482 (10 December 1870). The criticism intensified early in the following year, 520-1 (27 January 1871), 542-6 (22 February 1871).

77. Henry Cole, Diaries, 3 June 1870.

78. Devonshire Commission, vol. 1, qq. 22-8, 42-3, 49.

79. *Ibid.*, q. 54.

80. *Ibid.*, q. 5931.

81. *Ibid.*, q. 83.

82. *Ibid.*, qq. 85-98.

83. *Ibid.*, qq. 114-15, 117, 124, 127, 129.

84. *Ibid.*, qq. 232-3.

85. *Ibid.*, qq. 187, 201, 205.

86. *Ibid.*, qq. 220-4, 231.

87. *Ibid.*, qq. 192-8.

88. *Ibid.*, qq. 242-4.

89. *Ibid.*, qq. 779, 784.

90. Edward Frankland, *How to teach chemistry, being the substance of six lectures delivered at the Royal College of Chemistry in June 1872*, ed. G. Chaloner (London, 1875).

91. Devonshire Commission, vol. 1, qq. 788-91 (789).

92. *Ibid.*, q. 790.

93. *Ibid.*, qq. 803-4.

94. *Ibid.*, qq. 5690, 5693.

95. *Ibid.*, q. 5704.

96. *Ibid.*, q. 5720.

97. *Ibid.*, qq. 5728, 5767.

98. *Ibid.*, qq. 5789-91.

99. *Ibid.*, qq. 5894-5.

100. *Ibid.*, qq. 1210-11.

101. A.W. Williamson, *A plea for pure science, being the inaugural lecture at the opening of the Faculty of Science in University College London* (London, Taylor and Francis, 1870).

102. Devonshire Commission, vol.1., qq. 1190, 1204.

103. *Ibid.*, q. 1338.

104. *Ibid.*, q. 1364.

105. *Ibid.*, pp. vii-viii.

106. See n. 75 above.

Chapter six

1. D.S.L. Cardwell, *The organisation of science in England*, 2nd ed. (London, Heinemann, 1972), pp. 126-36. G.W. Roderick and M.D. Stephens, *Education and industry in the nineteenth century: The English disease?* (London, Longman, 1978), pp. 67-9.

2. 'Correspondence between the Science and Art Department and the Treasury as to the organization of the Normal School of Science and Royal School of Mines', *P.P.* 1881 [c. 3085] LXXIII. 563, pp. 32-5.

3. *Ibid.*, p. 12.

4. *Ibid.*, p. 13.

5. *Ibid.*, p. 18.

6. 'Copies of correspondence relating to the first report of the Devonshire Commission', *P.P.* 1871 [c. 422] LVI. 333, pp. 4-5.

7. *Ibid.*, p. 7.

8. 'Report of the Royal Commission on scientific instruction and the advancement of science', vol. 1, 'First, supplementary and second reports with minutes of evidence and appendices', *P.P.* 1872 [536] XXV.1, pp. ix-x.

9. *The Engineer*, 12 May 1871, p. 315. See also 28 April, 19 May and 2 June 1871 for further rhetoric.

10. T.G. Chambers, 'The Royal School of Mines and the Royal College of Science' in *Register of the Associates and Old Students of the Royal College of Chemistry, the Royal School of Mines, and the Royal College of Science with an historical introduction and biographical notices and portraits of past and present professors* (London, Hazell, Watson & Viney Ltd., 1896), p. xiii.

11. 'Correspondence as to the organization of the Normal School of Science and the Royal School of Mines', p. 33.

12. Chambers, 'Royal School of Mines and Royal College of Science', p. xliv.

13. Figures assembled from the annual reports of the Science and Art Department.

14. R.S. Lineham, *A directory of science, art and technical colleges, schools and teachers in the United Kingdom, including a brief review of educational movements from 1835 to 1895* (London, Chapman & Hall, 1895).

15. Harry Butterworth, 'The Science and Art Department, 1853-1900', PhD Dissertation, Oxford University, 1969; and J. Roach, *Public examinations in England, 1850-1900* (Cambridge, Cambridge University Press, 1971); for its replacement see D.S.L. Cardwell, *Organisation of science*, p. 210.

16. *J. Soc. Arts*, **20** (1872), 261-3, 725-35; *ibid.*, **21** (1873), 21-4.

17. D.S.L. Cardwell, *Organisation of science*, p. 132.

18. Charles Graham, 'Technical education: The introductory lecture delivered before the Faculties of Arts and Laws and of Science', 1 October 1879, University College London.

19. Michael Sanderson, *The universities and British industry, 1750-1850* (London, Routledge and Kegan Paul, 1972), p. 84.

20. A chemical technology degree course was introduced at Owens College in the 1880s, it included no mechanical engineering and had a heavy component of the principles of chemistry. 'Second report of the Royal

Commission on technical instruction', vol. 5, *P.P.* 1884 [c. 3991-iv] XXXI. (I), app. 33.

21. D.S.L. Cardwell, ed., *Artisan to graduate: Essays to commemorate the foundation in 1824 of the Manchester Mechanics' Institution, now in 1974 the University of Manchester Institute of Science and Technology* (Manchester, Manchester University Press, 1974), p. 149.

22. Sanderson, *Universities and British industry*, p. 85.

23. C.A. Russell, N.G. Coley, and G. K. Roberts, *Chemists by profession: The origins and rise of the Royal Institute of Chemistry* (Milton Keynes, Open University Press and Royal Insitute of Chemistry, 1977), pp. 87-8.

24. R.C. Chirnside and J.H. Hammence, *The 'Practising Chemists': A history of the Society for Analytical Chemistry, 1874-1974* (London, Society for Analytical Chemistry, 1974).

25. C.A.Russell, N.G. Coley and G.K. Roberts, *Chemists by profession*, pp. 114-15, 117-20.

26. *Ibid.*, p. 107.

27. James Dewar, 'Presidential address', in BAAS, *Reports* (1902), pp. 3-50. See the analysis of these figures in D.S.L. Cardwell, *Organisation of science*, p. 207.

28. Ministry of Education, *Education, 1900-1950* (London, HMSO, 1950); F.E. Foden, 'The National Certificate', *Vocational Aspect*, 3 (1951), 38-46.

29. C.A. Russell, N.G. Coley, and G.K. Roberts, *Chemists by profession*, pp. 270-1.

30. *J. Soc. Chem. Ind*, 50 (1931), p. 11.

31. On the Institution of Chemical Technologists see *Chem. News*, 101 (1915), 202-3, 211-14, 224-7; *Chem. Trades J.*, 60 (1917), 266.

32. On the Federal Council for Pure and Applied Science see H.E. Armstrong in *Chem. Trades J.*, 64 (1919), 95; and *J. Soc. Chem. Ind.*, 38R (1919), 18-19, 59.

33. On the Institution of Chemical Engineers see J-C. Guedon, 'Conceptual and industrial obstacles to the emergence of unit operations in Europe' in W.F. Furter, ed., *History of chemical engineering*, Advances in Chemistry Series 190 (Washington, D.C., ACS, 1980), pp. 45-75.

Index